LIVING
WITH
YARDS

LIVING
WITH
YARDS

NEGOTIATING NATURE
AND THE HABITS OF HOME

URSULA LANG

McGILL-QUEEN'S UNIVERSITY PRESS

Montreal & Kingston | London | Chicago

ISBN 978-0-2280-0856-9 (cloth)
ISBN 978-0-2280-0898-9 (paper)
ISBN 978-0-2280-0977-1 (ePDF)

Legal deposit first quarter 2022
Bibliothèque nationale du Québec

Printed in Canada on acid-free paper that is 100% ancient forest free
(100% post-consumer recycled), processed chlorine free

LIBRARY AND ARCHIVES CANADA CATALOGUING IN PUBLICATION

Title: Living with yards : negotiating nature and the habits of home / Ursula Lang.
Names: Lang, Ursula, author.
Description: Includes bibliographical references and index.
Identifiers: Canadiana (print) 20210300523 | Canadiana (ebook) 20210300981
    ISBN 9780228008569 (hardcover) | ISBN 9780228008989 (softcover)
    ISBN 9780228009771 (PDF)
Subjects: LCSH: Lawns. | LCSH: Lawns—Social aspects. | LCSH: Urban ecology
    (Sociology) | LCSH: Sociology, Urban. | LCSH: Urban ecological design.
    LCSH: City and town life.
Classification: LCC HT241 .L36 2022 | DDC 307.76—dc23

In one sense there is nothing more simple and more obvious than everyday life. How do people live? . . . In another sense nothing could be more superficial: it is banality, triviality, *repetitiveness*. And in yet another sense nothing could be more profound. It is existence and the "lived," revealed as they are before speculative thought has transcribed them: what must be changed and what is the hardest of all to change.

HENRI LEFEBVRE, *Critique of Everyday Life*, vol. 2

With love for my parents, Gretchen and Jeffrey Lang

With love for David, Anton, and Leonard

CONTENTS

What Yards Do

Yards are one kind of connective tissue.

Two women talk over a backyard fence, holding kids in clean white outfits on their hips. Hair tied back as they pause from their work. It is 1925, and these are Italian-American women in Minnesota. Photographs of this era seldom capture this kind of everyday moment. These moments that people make – and that, in turn, make people – can be invisible. These moments can be drudgery. Can convey inequity. Can bring the possibility of a new day. These labours can be a matter of care. The fence links, as it draws bounds.

Yards are human, and more than human.

In a backyard in St Paul, around 1920, three children stand amidst a flock of chickens. A woman stands partly out of frame, with a cap and long skirt, perhaps a domestic worker with a hint of the class and power relations circulating in this space. Chickens peck at the dirt. On a porch, around 1910, the girl holds onto a dog at the feet of a seated woman. She looks at the dog, the dog looks at the camera in the sunshine. These children all pause at their play. Fences, porches, rugs, tamped earth. Backdrops for living – work, rest, and play.

Yards are yardsticks, a measure.

Seen by experts as a reflection, a yardstick, a measure – of what? The Minne-apolis Housing and Redevelopment Authority captioned this photograph: "An example of extremely poor outdoor housekeeping and drainage prob-lems in a residential lot on Queen Avenue North, Minneapolis." The date is September 1960. The expert gaze sees health concerns in standing water and resident neglect manifested in an abandoned Christmas tree, detritus dotting the puddles. Have renters created this mess? Barrels overflow with trash, and a muddy expanse, rutted with tire tracks, leads to a garage in the background. The violence of urban renewal is coming, to be directed at people of colour and poor people. Yards are caught up in these visions. Potential vectors for disease and stagnation, to be improved. But really, as a reflection of residents' capacities of citizenship, of skill, for maintenance and cultivation. Reflections of a neighbourhood's value.

People maintain yards.

The two women complete their tasks. Apron in action, broom in motion. In 1910, in Minneapolis, Abby Foster reaches into her garden, pulling a weed or perhaps picking a flower. Her eyes squint as she surveys the garden before her. Across time, in 1979, another older woman also does a task outdoors, in the yard, near a house. She is unnamed and without a specific place. But still she sweeps the concrete pad, near a porch and steps. It is spring or fall – the leaves come or go, depending – and the sky is filled with branches. Her house is modest, and one decorative and worn filigree band stands upright while the other waits to be replaced or repaired. In both images, women embody these labours of maintenance and care. Bodies and tools, architecture and the out of doors, experience and task.

Bodies in the landscape and of the landscape.

She leans casually, a woman in the sun, one hand on her hip, the other arm along the top of the fence. Her smile competes with the stark white pickets for attention, echoed behind her in another fence. Herself a picket, upright contrasts with shadow, the shape of her arm at her hip interrupts the pace of light and dark. Her feet in dark shoes are hidden behind the grass. Property owned. Property bound. Her smile is infectious, in high relief because of the shadow across half her face. There is joy, pride, and a sense of ease. She is of this landscape of pickets – it is hers in a sense, and she is its. About 1910, Marie Madison King stands behind an outrageous shrub filled with blooms. She is almost entirely obscured. Just the length of her chequerboard skirt can be seen below, her face smiling but a bit sceptical. Maybe it's the sun. A glimpse of her chequerboard right shoulder just beyond the blooms. Both these women captured in a moment when their bodies are of their surroundings.

People come together in yards.

Special events mark the use of these yards, settings for one of life's great rituals. Furnishings are carried outside, dishes lined up on tables. Grass forms a carpet for the festivities. Two weddings separated in time and space. One, a wedding in North Minneapolis in 1985. In a front yard, people sit on folding chairs, smiling as the bride comes forth, her white high heel marking the way on the lawn. Thirty years before, a bride and groom sit with attendants, frothy dresses stand out against folding chairs. Together, they eat the celebratory feast under a fringed umbrella, rented for the occasion. People gather.

Yards measure rhythms.

People inhabit yards by sitting, reading, reflecting, thinking. Being out of doors. A similar affective energy circulates within these two images, made sixty-five years apart. A quiet calm pervades both. Women sit, on cushions, on an old wooden chair. A man reclines in a hammock, legs crossed, feet up. Positions to pause, to rest. In one image, books rest in laps at the moment when the readers look at the camera. It is 1899. The yard as a place of leisure. A place to listen, to smell, to touch. In the other image, a woman contemplates the camera, the person photographing her. A pause. The caption says "Life comes easy, June '64." Yet her expression seems to ask, is this life easy? The fence rolls gently with age and the sloping ground where dandelions grow. The yard as a place for pause, of ease, of rest.

A yard is made up of its inhabitants and what they do, how they live, the capacities they have. It can make us ask who has lived on these home grounds. How did they become home, and for whom? Yards can be places to pause, for time to stop. When do these pauses happen? Who has access to their rhythms? Yards are moments quick in passing, which happen again and again. Block after block, day after day, a yard grows, adapts, dies back, and regrows. Yards have become a part of the way people practise property. Again and again, property settles into a stable and bounded concept.

Relationships reverberate in the everyday experiences of yards.

This book asks, what relationships does a yard make possible? What can a yard do? What (else) might a yard become?

LIVING
WITH
YARDS

Living with Yards

This is where I *live.*

SANDRA in her yard, North Minneapolis

Yards are not quite wild, yet are rarely tamed. Yards are by definition out-doors, but are often closely associated with home interiors – they may be furnished and decorated. Yards are private property, but are by no means fully private. The state regulates yards at a variety of scales. Yards are spaces of human cultivation and care, but always in conjunction with the liveli-ness of non-human organisms and matter – which can be unpredictable, disruptive, and finicky. Yards are associated with residential landscapes of exclusion and privilege, even those whose inhabitants command modest financial means. Like other lands settled and redrawn by colonizers, yards have layers and layers of history underfoot. And yet, even with all these complexities, yards seem so common that we hardly see them.

Across diverse residential landscapes in North America and beyond, people cultivate front and backyards. In doing so, they participate in a form of everyday world-making, developing and expressing attachments and relations with their yards as inhabitants, property owners or renters, envi-ronmental stewards, perhaps commoners, and caretakers. *Living with Yards* takes an ethnographic approach to the ways people in one North Ameri-can city, Minneapolis, live with their yards and undertake diverse material engagements with them. The book shows yards to be at once physical, imag-inary, ideological, social, emotional, biological, and political places. This book uses the yard as a faceted lens through which to examine multiple and sometimes contradictory perspectives about how people live in urban environments and about how we collectively know those environments. Ultimately, *Living with Yards* aims to recognize and question the interplay between sedimented urban forms and habits, as well as the creative possibilities of everyday practice and experience.

The book also mobilizes yards as an analytic to better understand relationships between city policies and how people live with urban environments. *Living with Yards* explores what can be found out from the diverse ways yards are understood in urban environmental policies and projects, as well as how they are experienced and shaped in everyday life. Cities of all sizes are currently rethinking and integrating environmental management into public and private projects through concepts such as sustainability and resilience. And they are investing in policies and projects like green infrastructure. These adaptations of city habitats raise a host of issues. Of central concern is how everyday environments – settings for routines and habits, but also for creative emergences – embody both habitats and inhabitation. The book shows the ways habitats and inhabitation supplement and exceed one another, with an emphasis on critically assessing what a better understanding of urban inhabitation – in this case, how people live with yards – might offer and provoke.

What Can a Yard Do?

Walk along almost any residential block in Minneapolis, and you are surrounded by yards. Front yards stretch in both directions along a linear city block, with a concrete sidewalk and usually a narrow boulevard. The block might be punctuated by driveways, or back alleys might provide access for cars. The northern continental climate makes for lush and green short summers, brown autumns, and snowy, long winters. Along the rectilinear streets first built along streetcar lines in the 1910s and 1920s, yards are very often similarly shaped – long lots with a single-family house set back from the sidewalk and situated slightly to the north of the property to allow for a sunny side yard to the south. The financial value and size of homes vary significantly across the city, as does the presence of a mature urban forest canopy, and the proportions of renters to owners. Although the yards are similar to one another, somehow it's easy to see where the property lines are drawn. Plants make boundaries, just as edges of mowing do, or the degree to which lawns are manicured – weeded, lush from fertilizers, littered with fallen leaves. But in most cases, these spaces provide a connective green space along blocks and neighbourhoods. Yards constitute one kind of utterly ordinary urban space in many North American towns, cities, and suburbs. As a result, the life of yards is often overlooked, simplified, or reduced.

Yards can be simple and obvious. They are the outdoor space around homes. Grass grows there, sometimes trees and other plants. A general observer naturalizes these spaces and overlooks their particularities. Or, through yards, people can display and reproduce everyday banality and triviality – the cheap nylon flag with a symbol of the season or uniform hostas framing the bounds of private property, because that's what a homeowner does. These are yards filled with consumerism and a kind of market-induced predictability. Or maybe yards can be habitats, shaped and planned by market forces, expert knowledge, uneven power, and state control. These are gridded parcels of private property defined by city streets and alleys. Originally laid out across the land in the dreams of developers, now these yards are regulated by an apparatus of city codes and enforcement defining the setbacks, the contents, and the types and extents of plants, and assessing the maintenance of screens, doors, paint, and address numbers. Increasingly, though some argue not fast enough, shifting ideas about climate change lead to shifts in governance. Inhabitants don't always see the structures that profoundly shape the possibilities of everyday encounters with such spaces.

Somewhat surprisingly, scholars in urban studies, design, and geography have often overlooked yards. The majority of scholarly work on yard and garden spaces around homes has come from the fields of landscape architecture history, cultural landscape studies, and architecture history. These approaches look primarily at the history and meaning of designed gardens, lawns, and turfgrass. Such studies have often pointed out the constructed nature of landscapes, and the ways ideologies of labour and capital become inscribed in particular landscapes – usually focused on the development of green turf on eighteenth- and nineteenth-century British country estates, and the importation of miniaturized pastoral ideals into suburban American landscapes.[1] Vernacular residential landscapes in the United

States have generally not received the same scrutiny or attention.[2] When they do, the focus often remains on the phenomenon of *the lawn* (often understood through popular culture references and caricatures), rather than a more inclusive look at all the different materials, activities, and meanings of yards as residents actually practise them.[3] In these accounts, the lawn stands in for nature. The lawn is understood to be contemplated, tamed, nurtured, and cultivated by people, much like a raw material to be given shape and form. Or the tendency is to see spaces such as yards as static realms reflecting aesthetics of designers, mass-market trends, and the consumerism of the burgeoning postwar middle class. Sometimes yards are understood through regional histories of vernacular forms.[4] Or particular yards warrant interest as folk art or the expression of entirely individual idiosyncrasies.[5] Most recently, yards are understood to be a realm of environmental decision-making, with the presence or absence of lawn, food plants, or native species indicating a person's environmental values, and sense of self in relation to these.[6]

But look also for yards as places of lived experience and the making of meaning, where bodies and landscapes interact through practice and affective attachments. This is the yard where someone bends over a garden bed, reaching to pull weeds as they have done a hundred times before, because they want to cultivate sensations like blooming plants or different textures for neighbours to see and experience. This is a yard where women might stop to talk over a blooming plant, and those blooms bring back a past for one of them. Or maybe this is the yard that shows to others the inhabitant can't maintain their home, through the scruffy grass, the garden beds overgrown, trees that need trimming. In this yard, maybe there will be a city fine to pay for letting things go. Maybe the inhabitant struggles to pay the mortgage. All of these are inhabited yards, constantly made and made meaningful, physically shaped and reshaped. These yards constitute geographies of home, over time.[7] Places where different organisms, histories, skills, and bodies all live together as interwoven ecologies over time.

Really seeing these familiar, ordinary spaces will help us address some of the most fundamental questions facing urban environments. How do people make meaningful lives together in proximity to difference? What kinds of relations exist, and might be imagined, between people and their more than human surroundings? Who has access to a space like a yard, or to the relations of cultivation, care, and inhabitation that often happen in yards? How might we expand such access to those relations, even as cities become more dense and yards disappear? As the book unfolds, I argue that yards serve as a useful analytic through which to see some of these critical relationships and forces at work in urban environments, including contemporary urban environmentalisms organized around concepts of sustainability and resilience.

Urban Environmental Politics in and through Yards

Contemporary urban environmental politics increasingly enrols the yard as a crucial interface through which to rethink challenges, including relationships among water, city infrastructure, native species, chemical pesticides and fertilizers, and local food production. Municipal policymakers and planners, as well as environmental advocates, have tried to rethink urban environmentalism beyond green spaces such as parks, and in terms of sustainability and resilience. They have implemented policies and projects to move regulations and voluntary changes in behaviour toward quantifiable metrics and goals. Such initiatives imagine the city as a series of discrete and manageable biophysical systems and surfaces. These diverse efforts reduce complex social processes to measures and management, and, in doing so, often miss the opportunity to engage with the everyday politics of difference that affects how urban environments are made, and made meaningful.

Sustainability policies and projects operate within the context of increasing pressures on cities and towns to attract capital investment and to maintain aging infrastructure with smaller and smaller budgets. Greening projects often are thoroughly capitalist development projects, reinforcing and exacerbating inequities across cities in the context of neoliberal urbanization.[8] This kind of urbanization emphasizes individuals, their health, well-being, and bodies. And it imagines the individual – most often understood as a homeowner with an individual property – as a rational actor and as an environmental consumer. Municipal governments seek to mitigate negative environmental effects at the scale of the individual owner and the individual property parcel by emphasizing and monitoring attitudes, behaviours, and choices. Such governance perspectives imagine yards as an aggregate of self-contained units conforming to property bounds.

Increasingly, various sectors organize urban environmentalism around the concept of resilience. This grammar focuses on a city's capacity to "survive, adapt and grow" in response to ongoing "chronic stresses" and unpredictable or unexpected jolts of "acute shocks."[9] Sustainability language implies that there might be a more harmonious relationship between nature and city. In contrast, resilience suggests that environmental calamity and breakdown is inevitable and ongoing. Its purview, then, is how we might collectively handle – and even benefit from – these inevitabilities through planning for "proper" urban growth and density. Resilience is imagined as possible at all, or multiple, scales, ranging from an individual to a neighbourhood, city, state, and nation. Resilience is also thoroughly embedded within neoliberal vocabularies of value, ownership, and imaginaries of the market as being outside social relations.[10] Resilience, like sustainability, involves all the usual suspects – private, public, and nonprofit sectors – partnering toward generating maximum "resilience dividends." Several lines of critique have argued that, like sustainability, resilience is a flexible concept that can encompass everything, and consequently mean almost nothing. Critics have argued that resilience is often deployed as an apolitical term, implying that capacities to respond to long- or short-term challenges are evenly distributed and available equally across social differences and uneven geographies.[11] Furthermore, they argue, resilience places burdens of adaptation and restructuring on individuals and often marginalized communities, ensuring that systems like global capitalist urban development continue – and creating mismatched clashes of power and influence.

The circulation and promotion of urban sustainability and resilience policies and best practices that aim to reshape spaces and surfaces such as yards emphasize the making of habitats as environmental management problems with physical solutions. But yards also constantly shape and are shaped by financial and legal structures, the risks and benefits of particular infrastructural geographies, ever-changing biophysical habitats and ecosystems, and settler colonial processes of dispossession, urbanization, and development through exploiting social difference – such as persistent racial segregation in housing. As explored throughout the book, we must supplement how we approach yards, beyond the technosolutions and "biological solutioneering" of much of the conversations happening within sustainability, resilience, and adaptation planning discourses. This book does not discount or ignore the serious ecological consequences (and possibilities) of specific yard practices, whether more conventional, such as growing turf-grass lawns, or more alternative – for instance, cultivating rain gardens or native habitats. But my focus remains on social and cultural dimensions of yard practices and spatial processes of yard making and maintenance. Most broadly, we must ask how looking at urban environments primarily through the prism of individual ownership or biophysical processes shapes our collective capacities to reinforce and maintain particular socio-spatial relations. How do people understand these capacities? And how might we differently imagine and enact a more inclusive sense of socio-environmental well-being?

Changing Urban Imperatives and Habitats

Contemporary municipalities are under enormous pressures to manage and maintain various public services and imagine new futures within the neoliberal context of shrinking budgets and the undermining of shared striving for a broader public good. These ideals have always been flawed, racialized, and partial, laden with uneven power and privilege. Furthermore, city budgets dependent on property values reeled after the 2008 financial crisis. In addition, increasingly severe and numerous weather-related events such as floods and fires put pressure on urban environmental planning and management. Municipalities are in the midst of trying to rethink environment and environmentalisms beyond green space, in much more pervasive and fundamental ways. How do these changes and adaptations take shape, in response to the complex imperatives cities face?

Cities have been reworking their physical nature. This process begins with understanding cities as habitats – thinking about all the expertly planned, designed, built, and maintained physical settings of urban life – for example, upgrades to aging infrastructure for water and transportation, urban design projects to rework shared open spaces, or architectural responses to increasing density. Henri Lefebvre identifies a chasm between habitat and inhabitation, and that awareness underpins this book. *Habitat* encompasses specialized knowledge, rendering the city as a knowable and definitive object. Thus, for fields such as urban planning and engineering, the city becomes a series of isolated functions and territories with fixed, even if often overlapping, boundaries. On the opposite end of the spectrum, *inhabitation* entails how people live with, through, and against those habitats, aspects of which I show with this book. Advancing this distinction further, my central question is how (and what) can we learn from how people live?

If rethinking green urbanisms beyond green space is one significant contemporary urban imperative, austerity and capitalist urban (re)development are others. Austerity in the form of severe public-sector funding cuts has exacerbated ongoing processes of neoliberal urbanization. The impact of these cuts takes shape within the longer term hollowing out public goods, promotion of public-private partnerships and entrepreneurial imperatives to attract capital investment, and shifting of responsibilities from public-sector institutions onto individuals and nonprofit organizations. Beneath all lies the land,[12] and the perceived need for continuous capitalist growth and development means city habitats are constantly changing. These changes occur at different rates and with particular geographies, underpinned by reliance on inequalities and social differences such as those connected to race, gender, sexuality, class, and more. Yards and their configurations in daily life register the processes of ongoing racial segregation in housing and disinvested neighbourhoods, as well as the voracious movement of gentrifying capital investment. Green urbanisms have been a realm where these forces converge. The greening of urban habitats is a response to and continuation of these converging urban pressures.

Everyday Life Might Be Otherwise

All the while, cities thrum with people's everyday practices. How people live shapes, and is shaped by, everyday habitats. Although daily life often feels sedimented and routine, it is shaped again and again by ongoing processes of making – the production of space and society. There is a dual nature to everyday life. On the one hand, the everyday is a repetition that is dull and predictable, the same thing happening again and again. This is an everyday dominated by capitalist linear time, the demands of work and ownership, the drudgery of maintenance. On the other hand, this repetition makes it possible for difference to emerge – something different *could* happen tomorrow. If only we might see the potential for difference in this repetition, we might see a less settled everyday.[13] This daily life is interwoven with power-laden structures and relations. These shape what becomes mundane and overlooked, utterly ordinary. And yet this everyday is still to be determined. This is an everyday that might be otherwise.[14]

There is so much variation in everyday life we do not see – in form, use, and meaning. For such familiar and ordinary spaces, there are a surprising number of things happening in yards. Some feminist political economists urge broadening narrow understandings of "the economy" as something external to society, as a machine with levers, as an objective realm filled with neutral and rational decisions. Rather, community economies of surviving and thriving contain much more expansive diversity and possibility than dominant concepts and theories take into account.[15] People share resources, care for others, trade and swap, mend and sew. Spaces like yards are filled with diverse practices, meanings, and physical features that can be understood in similar ways. This more expansive view of yards points to the need for much more diverse theorizations of concepts – such as *property* – that shape the possibilities of everyday life.

People practise property in and through yards. Caricatured as the moat of everyman's castle, yards have been considered a bastion of what geographer Nick Blomley calls the ownership model of property.[16] And legally, in a sense, they are this. A territory is bounded by survey lines drawn as part of an apparatus of historical development and regulation, valued by financial systems, and understood to be property. This ownership model of property continues to shape access to spaces like yards. But, property is practised day in and day out, and the practice includes how people reinforce or break down such bounds. It is an idea made manifest through the ephemera of survey markings on a map and in the corners of urban lots. As such, it can be reinforced or disrupted each and every day by what people do with it. Because homeownership is so tied to financial security and the accumulation of wealth in the United States, such ownership also shapes the distribution of resources and capacities for particular yard practices. Thus, it is crucial to better understand private property ownership in urban North America, and to see these spaces in relation to geographies of poverty and affluence.[17]

Yards may belong to people as owners, but human inhabitants also come to belong to yards. What is the potential of this kind of belonging?[18] If we shift the ways we think about what property is, what it means, and how it is valued, yards can provide new ways to understand urban histories of dis-possession and settlement, inhabitation, and cultivation. Conceptualizing yards in this way is steeped in feminist political economy traditions that have sought to rethink economic concepts such as work, along with geographies of social reproduction, care, and domestic life.[19] If we understand more about how the meanings, practices, and spaces of everyday life become so settled in common sense ideas like property or narrow modes of gardening, perhaps we might identify some further ways to unsettle and remake the world. However more or less immediate, and more or less wholesale in nature, this is the perennial motivation of critical geographic approaches.

Affective Force and Affective Ecologies

Geographies have come alive as relational assemblages of lively materials, distributed agencies, and animated political and material flows. The variously named and nuanced material turns in the study of nature-society relations – whether called more-than-human, posthuman, or vitalist – have been useful ways of animating the material world and engaging with non-human organisms (like plants and animals) as socio-politically meaningful.[20] Yards are more than human. But pointing out that fact is not quite enough as we think about how yards participate in the making of human-social relations, because yards are also deeply human. People are affected by the more-than-human organisms of yards, just as these organisms and surroundings are affected by human praxis. Ecologies unfold from this affective force between people and surroundings. Yards have their own kinds of affective force in people's individual and collective lives. Yards are deeply intimate and interior spaces for people. They are spaces that have highly individualized meanings, memories, and practices, even as they constitute connective tissues in the shared landscape where differences are negotiated. Yards are sites of material engagements with nature, and they are also deeply felt emotional realms. These social geographies of self and surroundings are made through interstitial and circulating atmospheres.[21]

Phenomenology is useful as a guide here, because it is a broad approach that sees the affective force of objects (or surroundings) in interaction with processes of subject formation. How a subject, or sense of self, comes into being depends on a back-and-forth encounter with its object, or surroundings. French philosopher Maurice Merleau-Ponty described this encounter as a deeply fleshy process, always between the flesh of the world and the flesh of the body, dependent on processes of perception.[22] If we consider yards in this way, these spaces interrupt, invite, reinforce, demand, delight, frustrate, challenge, and surprise human inhabitants through biophysical, material, and sociocultural processes. This takes shape in embodied practices. All the digging, dragging, carrying, lifting. Hands in soil, dirt under fingernails. But also in the sitting and the listening. Looking out of windows. Attuning perception to changes in leaf colour, bloom times, a neighbour's way of mulching. Noticing a neighbour's neglect or maintenance. These capacities, skills, and perceptions in and through the human body, in relation to other non-human bodies, are shaped by life histories and past experiences in place. So, these capacities are also always shaped by social relations and social differences.

Through yards, we can see new layers of attention to perception, sense, and feeling added to approaches emphasizing the world as more than human. Environmental care is a realm where these layers add up. There is the potential for pleasure and joy in environmental labours, especially in the work of conservation and environmental management, and the ways in which environmental subjectivities are constantly in the process of becoming through practices of cultivation and maintenance. Such practices can be highly gendered and unequal, and can entail burdens. These same practices can also be sources of new ways of being and becoming in the world. The development of subjectivities, or senses of self, is not separate from these earthly matters.[23] In fact, this co-becoming with the world and others involves ongoing and inseparable relations of humans and earth. Dominant modes of property, environmental management, and oppression narrow these co-becomings. But such emergent ways of being in common with others nevertheless survive, resist, and persist.[24] Even in such mundane spaces as yards, how we understand and enact what seem to be individual subjectivities directly impacts what becomes possible and what becomes in the world, as well as with whom. Aggregates of these subjectivities become meaningful through sheer repetition and quantity. But they also add up to more than a sum of parts, with possibilities for reworking and transforming oppressive structures into new, more collective co-becomings, and their entangled ecologies.

Life histories, accumulated knowledges, and experiences are at the heart of environmental care labours such as gardening and cultivation, and also the affective ecologies emerging from simply being outdoors. The thing about gardening and maintenance is its accessibility, and universality, over space and time. It is a practice not limited by education level, age, or income. There is always a way to make a garden work in the most unlikely places and times. To try it out, to cobble it together, to share in it, to take it apart and rebuild it. This accessibility is powerful because of its ubiquity, its ordinariness, its taken-for-granted-ness. These are relations of ordinary affect and emotion, power and experience. Although it only makes sense that gardens and gardening are associated with yards, I insist that yards are more than gardens, gardening, or residents choosing which plants to cultivate. If we see yards only in this way, we might decouple these specific practices and spaces from meanings. As North American cities become more dense, yard territories may be lost. But perhaps the relations that have been at the heart of these spaces could be reworked and renewed – potentially in more collective, shared, and socially transformative ways.

Phenomenal ecologies are sensed and felt combinations of material surroundings, everyday practices, and the attachments (or lack thereof) between bodies. To try to recognize the phenomenal ecologies of yards is to acknowledge – even if it's challenging to grasp within representational strategies like writing, storytelling, or image-making – that yards are felt realms. They are felt through intentional practices of design and cultivation, through the drudgery or pleasure of maintenance and keeping chaos at bay, and through simply being outside and feeling a breeze or sitting on a stoop. Yards are felt in the breath – green spaces breathing for the city. They are felt in the interplay between the regulatory apparatuses shaping space and what people do in these spaces. Yards are also felt realms in the ways people notice and negotiate difference.

Rhythms and Time Are Essential to Understanding Yards

All the preceding aspects of yards can be seen through experiences of repetition, time, and rhythm. Rhythm brings together aspects of time and repetition, made up of cyclical repetitions and linear progressions of passing time. Henri Lefebvre uses an analysis of rhythm to get at the processes of commodification and alienation integral to modern capitalist production and the everyday experiences of people caught up in these processes. The introduction of modern linear time to daily life disrupts and displaces the cyclical repetition of nature, according to Lefebvre. But fragments of natural time also hold on, return, fade, and grow stronger. This polyrhythmicity – the simultaneity and interference between linear capitalist time and more cyclical recurring repetitions of natural life – requires a particular kind of listening, or what Lefebvre calls rhythmanalysis.[25] This is a listening for houses, trees, wind, stones, bodies. Lefebvre brings together temporalities and spaces through rhythms, reconstructing the separation and alienation of the two that has emerged with modern capitalism.

Homeownership means that people live with yards for long periods, often over decades, of their lives. These long relationships between human inhabitants and yards are marked by non-human organisms and biophysical processes – for example, by the growth and decline of long-lived plants such as trees. Changing family configurations also affect the ways people relate to yards. Younger children venture further afield as they get older; older children grow up and move out of the house. Older adults downsize or become less physically able to keep up with home and yard maintenance. All these changes to do with the arc of life histories affect yards. People grow with their yards, and vice versa.[26] Rhythms also have a material accretion. Materials build up. Window trim is painted and painted again. Gardens mature and plants thicken. Yards and houses are composed of layers of modification, repairs, and renovations over time. Such iterative actions and practices produce an accumulation and accretion to the rhythms of how we live.

Yards can be in and out of time. Cycles such as changing season happen in yards over relatively long periods. Yards express this phenology through changes in plants and in animal behaviours, as well as associated human uses of houses, porches, and decks. Attunement to seasonal changes can be sources of comfort, require particular expertise, or disrupt the usual flow of time by demanding certain labours (e.g., weeding, mulching, trimming). Changing climate has meant changes to the usual timing of seasons, and noticing this disruption can be a source of worry, grief, or pleasure. Yards can also be a means to step outside perceptions of linear time altogether – for instance, in getting lost in tasks of the yard, like weeding, or in travelling in memory times, through the smell of this plant or that way to stake raspberries. Such excursions are often understood as welcome relief from harried schedules.

Who has access close at hand to these rhythms and times-outside-of-time? Inhabitation is spending time, being with others, being with self. Given that so many of these engagements over long periods have to do with homeownership, and that homeownership is a key indicator of broad and racialized economic inequality, there are crucial dimensions of access to consider. If such relations associated with inhabitation are forged at times effortlessly through the pleasures and labours of yard experiences, are there other realms where such relations can take shape, less constrained by the limits of the private-ownership model of property?

Knowing Urban Environments: Stories and Visual Ethnography

Inviting and Listening to Stories in Place

When we explain the world, we make the world, in a way. Embedded in our stories are future possibilities, and future worlds. The stories people tell make sense of lived experiences, order the world – even for a moment – and communicate shared pasts. Dominant modes of knowledge – be they geographic, environmental, or policy oriented – too often discount, erase, and overlook stories told by people from marginalized groups. Feminist geographers and others highlight taking stories seriously – something increasingly appreciated in the context of contemporary environmental change. These research traditions within feminist geographic methods, ethnography, and environmental humanities approaches inform my research.[27] I aim to take seriously what people already do in the ordinary and in their own everyday lives. I listen to stories in place about how people live with yards. I want to expand the kinds of environmental knowledge we might collectively value and the ways we might bring socially differentiated environmental experiences into our decision making.

This requires recognizing research praxis as always situated, partial, and unable to fully enclose urban environments and experiences within neat categories. Feminist and queer geographies reclaim what dominant scientific perspectives view as "limitations." These partialities become essential invitations to a more reflexive research praxis – one more aware of power-ridden relations between researcher and researched. The identities and capacities of researchers can be sites of examination and reflection. Because I conducted this research entirely on my own, I was limited in scope and access by my own capacities and identity as a thirty-something, white, able-bodied, cisgender woman. I was also a new mother, novice gardener, and primary English speaker. Other researchers would have different relationships to yards and participants, enabling many valuable and necessary contributions to yard study. This more reflexive research praxis enables experimentation and analysis through registers of writing and visual representation, as well as involving people in analyses of their own experiences. Ethnography allows me to learn what residents say about yard spaces, in relation to what they actually do, make, and feel in these spaces. This approach allows me to be in others' time and space.

The interwoven nature of embodied practices, meanings, histories, and physical surroundings remains a methodological challenge. Studying yards needs to take into account registers that exceed the familiar representations that might emerge in stories and spoken descriptions, because a lot of what happens in and through yards has to do with how people feel in yards.[28] In other words, yard study needs to encompass all the affects and emotions of environmental care, embodied engagements with more-than-human worlds, navigating social differences and relations with others, and the rhythms and temporalities of all these attachments.

"Show Me Your Yard"

I examine these issues through a detailed case study approach, focusing on how people practise and experience residential yards in three neighbourhoods in Minneapolis. Using multiple methods centred around ethnography, I offer a textured analysis of people and their home environments. I also examine sustainability through analysing municipal policies, metrics, and reports. I draw on interviews I held with planners, policymakers, and environmental managers. In order to understand opportunities and challenges in environmental advocacy relevant to yards, I conducted participant observation and interviews with landscape designers at a nonprofit organization that promotes residential rain gardens throughout the metropolitan area. For general background about the yard and gardening scene in different areas of Minneapolis, I participated in a range of local events about yards and gardening – for instance, gardening fairs and neighbourhood garden tours, sustainability events and local conferences, and rain-garden information sessions and volunteer workshops. Throughout all of this research, I linked scales of body, home, city block, neighbourhood and city through the study of embodied yard experiences and practices.

The centre of my research is visiting yards and involving participants in several primary ways. First and foremost, I conducted in-depth yard visits with residents in approximately forty-five yards in three study areas, following ethnographic methods developed in my pilot study as well as by Head and Muir, and Arnold and Lang.[29] These yard visits comprised primarily semi-structured and unstructured interviews during yard tours led by residents, along with participant observation in the yard, with residents when possible.

I started each yard visit with basic background questions about the yard, house, and household, as an opportunity to establish rapport with participants and to ensure consistent data to characterize yards/homes for analysis. Then, I asked residents to show me their yards. As we walked around the yard, I looked at both its physical characteristics – physical boundaries, species that inhabit the yards (especially plants and trees), structures such as houses and garages, connections to neighbouring properties) – as well as how people talked about and inhabited these spaces. I documented the yard visit as we walked and talked – taking photographs and recording the conversation for transcription, and making detailed field notes immediately following the yard visit. I approach photography as a reflexive research praxis, requiring and inviting awareness and reflection about how the making of photographs impacts relations with participants and yards. Documentation of the outdoor spaces is a major component of the project, including photographs, drawings, and diagrams made by myself and sometimes with participants. Over five thousand photographs were made, edited, coded, read and reread, pored over for details missed.[30] Throughout the writing of this book, photographs served as portals into sensory memories of weather and seasons, how bodies moved and paused in yards, and conversations made from topics circling the material at hand in plants and trees, from intimate inward and outward reflections between inhabitant and surroundings.

Study Sites and Minneapolis

Minneapolis is a generative place in which to study the intersection of urban sustainability policies and residential practices due to its urban morphology, neighbourhood planning tradition, and relatively forward-thinking sustainability efforts. Minneapolis is known for innovative approaches to incorporating community-led planning into the heart of planning processes, from neighbourhood to regional scales.[31] The city's population is about four hundred thousand, and the total population of the Twin Cities region of Minneapolis and St Paul and the surrounding suburbs is about three million. In their urban and suburban forms, residential landscapes within Minneapolis are fairly typical of those throughout much of the midwestern and northern United States. The outer edges of the city grew in the 1910s and 1920s as a network of streetcar suburbs, laid out with rectilinear streets and alleys. Minneapolis's median density single-family housing is largely made up of rectangular lots about forty feet wide by one hundred feet deep, with house, yard, and garage. This common urban fabric now stretches across neighbourhoods with varying socio-economic and demographic characteristics.

This study focuses on three sites within Minneapolis, where I became acquainted with more than forty-five yards and their inhabitants. In Northeast Minneapolis, the historically white working-class neighbourhood is organized around railroad lines and related industries, and comprises small single-family homes built in the 1920s. Similar to other more affordable areas within the region, its increasingly diverse population includes immigrants originally from Central America, Africa, and Asia. The study site in North Minneapolis was originally developed in the 1920s with many large four- to six-bedroom homes on spacious lots, which housed middle-class and upper-middle-class Jewish residents, followed by African-American residents as Jewish families moved into western suburbs during the 1960s. The study site in South Minneapolis is primarily a white upper-middle-class area with homes built in the 1910s, near several of the famous lakes in the city. Redlining and racial segregation have strongly affected residential development in Minneapolis, and their effects persist in a highly racially segregated Minneapolis today.[32]

Though such neighbourhoods were originally imagined as suburban, inhabitants now understand them as thoroughly urban in relation to suburban and exurban places. As with almost all residential neighbour-hoods in Minneapolis, except those in the densest downtown areas, these neighbourhoods today are still composed mainly of single-family homes or duplexes with yards. The yards reflect how inhabitants have refashioned, to varying degrees, these once-suburban landscapes in the intervening decades. The inhabitants who participated in this study differ from one another in terms of class, race, employment, age, and length of time liv-ing in their homes. Participants were nearly all homeowners. The yards range from bare and self-described "neglected" spaces of primarily grass to intensely cultivated and densely gardened worlds. For the most part, par-ticipants did most yard work and gardening themselves. If they did hire or rely on outside help, it was mainly for basic lawn-mowing services.[33]

A range of design and scholarly perspectives imagine and represent yards as singular territories separated from their socio-spatial surroundings, rath-er than as connective – and even collective – urban and suburban tissues. Popular home and gardening magazines usually represent images within one yard, obscuring neighbouring yards by framing and focus. Or design approaches celebrate associated architectural features such as front porches, as facilitating connections with the social life of the sidewalk and street. In scholarly studies, yards become a realm of individual decision-making, or are studied primarily in terms of residents' decisions about gardening or lawns. To better understand how yards might be functioning as connective urban tissues, and in relation to other neighbouring yards, I focused on between five and ten adjacent city blocks in each of the three study areas. In contrast to a more singular approach to yards, I became familiar with as many yards as I could in each study area. This included all types of cul-tivation, levels of maintenance, and physical features in yards. I conducted full yard visits in approximately fifteen yards in each of the three study areas. This enabled me to become familiar with the relationships between yards at a relatively fine-grained resolution and in conjunction with phys-ical proximities, rhythms of habit, and social connections. This territorial approach, which flattened yards to an area on a map, enabled familiarity with the complexities of interwoven social and biophysical relationships, in many cases crossing and exceeding property bounds.

Since 2003, the City of Minneapolis has implemented a range of city-wide sustainability-related initiatives. Originally framed through global discourses of development, *sustainability* has been understood as integrating the "three E's" (environment, equity, economy). In Minneapolis, a focus on sustainability was part of a second wave of urban sustainability efforts across North America, integrating goals into policies and city business across existing departments. Minneapolis has continued to work hard to brand itself as a "green city," and it emphasizes an extended history of concern for the environment in this branding.[34] The city's sustainability agenda intersects with yards in several ways: water quality improvement goals; native wildlife habitat, urban forest management, and local food production. In addition to these formal activities on the part of the City of Minneapolis, there is an active and growing world of gardening promoted by nonprofit organizations, community groups, and for-profit businesses – some of whose activities overlap with yards and yard practices.

Living with Yards

Together, these aspects of yards add up to a configuration of living with others. Living with surroundings and others (human and non-human) ranges from purposeful cultivation, maintenance, and care, to simply being present with others through relations of response and attunement. To make sense of these relations, I draw from a constellation of related concepts – none of which on their own quite captures the experience, practice, and co-becomings of living with surroundings and others over time in spaces such as yards.

The spaces and practices bleed across the whole spectrum from habitat to inhabitation, but there are differences in emphasis in the ways we design, manage, and critique everyday environments. The approach of this book is a relational understanding, building on dwelling perspectives.[35] This approach foregrounds an orientation toward bodies understood as engaged in their surroundings through materialities, praxis, and affective ecologies of co-becomings. Yard rhythms and temporalities make the lived experiences of maintenance a site for both creative possibility as well as the drudgery dismissed by dominant understandings.[36] The heart of the argument of the book is that we consider *living with* as a way to describe this range of activities and experiences most meaningful in the everyday, and that we consider the implications for reimagining spaces like yards through an emphasis on the inhabitation end of the spectrum.

The yard visits of this project produced a rich array of yard experiences and spaces. I organize yard stories along this spectrum of yard engagements, ranging from the most intentional and skilled practices of cultivation as gardening, to the broadest practices of sitting, reflecting, and being attuned to senses out of doors. Not always positive or celebratory, yard stories capture interwoven sets of relations. I argue that we must think beyond the city as habitat and consider how spaces such as yards are inhabited and meaningful to inhabitants. Doing so requires understanding how bodies engage with surroundings through everyday practice, how this engagement unfolds over time, and what kinds of socially transformative possibilities might be latent within this *living with* these everyday landscapes. These understandings entail examining encounters with surroundings as ongoing formation, shaped by socio-natural rhythms. What might these insights offer our understandings of property, environmentalism, and experience? The possibilities of these encounters – both materially and socially – circulate through these embodied engagements. And these encounters might be reworked if we see these engagements in terms of socio-spatial relationships with the potential to be reconfigured.

Structure of This Book

The arc of this book extends from dominant and abstracted visions of yards into the intimacies of yard experiences, ending up at some of the possibilities and limitations of the ways people practise property, and might practise property differently. Throughout, I aim to show how individuals are always already social beings, part of power-laden relations of difference and negotiation circulating in and through everyday environments like yards. This recognition is important because how we understand what is "social" shapes our potential capacities to imagine collective life. It throws into question neat and binary notions of community/individual and public/private. In practice, these notions are much less rigid and less easily defined. Governance and politics are not easily contained in particular categories and are always mediating and mediated by experiences and understandings that develop over time. Sociality is always a process that is ongoing and full of continuities (the "second natures" of habit and routine) and ruptures (the aspirations, creative desires, unexpected failures, and limits). This sense of being social is always intermediated with environment and surroundings, in large part through embodied experiences – especially in spaces of cultivation and inhabitation like yards. Urban terrains offer particular social possibilities, as do living things, for how we understand and inhabit them.

Recent iterations of dominant techno-practices claim to dream and plan the city yet to come as sustainable, resilient, and adaptive. Tracing such iterations in Minneapolis, chapter 1, "Building Urban Habitats," asks how yards figure in these emerging green urbanisms. The chapter looks at the shifting purposes and practices of gardening as it intersects with yards. As city codes transpose inhabitants' doings into their material effects, inhabitation itself is largely written out of official imaginaries of future green urbanisms. At the same time, these official visions are only one part of a complex terrain of green urbanism, involving much broader practices and projects than are usually considered in either official plans or critiques that rely primarily on those plans.

Ways of knowing everyday urban environments through inhabitants' practices and experiences entail highly intimate, creative, complex, and often communal capacities. A residential yard is one place that shapes, and is shaped by, these capacities and experiences. In the book's second chapter, "Rhythms of Inhabitation," I make sense out of the diverse array of yard experiences and spaces. As I weave yard stories with conceptualizations of inhabitation and cultivation, these familiar spaces become important lenses into worlds of everyday life and practice. This chapter tells stories of everyday environmental care labours, creativity, drudgery, and maintenance, as well as being with others and with surroundings over time. If expertly planned and measured habitats anchor one end of the spectrum of urban environmental experience, then the other end is inhabitation. Along the way, are modes of making, cultivation, and maintenance that involve more or less creative and everyday environmental care.

The third chapter, "Chasing Yard Affects," explores how yards are felt. The chapter responds in the form of a photo essay to the analytical and methodological challenges of how to study, write, and convey affective capacities within the everyday life of yards. Image and text together embody the proposition that affective dimensions of yard experiences constitute the heart of what people do in their yards, and how they understand these spaces. For the most part, yard affects happen in registers beyond more easily legible concerns such as property value or environmental outputs. This chapter is a photographic and textual experiment in evoking some of the range, virtuosity, intimacy, feeling, and encounters in yards in the study. Yards make possible the circulation of particular affects. Yards are, at the same time, made possible by those circulations.

The fourth chapter, "Practising Property and Life in Common," examines entanglements between property, everyday practices, and common urban life. Beyond yards as spaces of private meaning and experience, there is a meaningful life of yards as commons and spaces for commoning. Shared yards, informally communally gardened yards, and the circulation of plants and knowledge all constitute aspects of this common life of yards. In and through yards, commons and commoning can be found interwoven with the logics of private property. These spaces make possible one way to see commons in tangible terms, through which variations in spatialities and temporalities – as well as the role of non-humans – add to emerging conceptions of urban commons and sustainability. Such practices are not equally available to all, and commoning in and through these small parcels raises questions at an intimate scale about broader efforts to redress and repair dispossession of land, wealth, and labour.

To conclude the book, I discuss how inhabitants' experiences with and desires for certain environments match up or conflict with urban environmental visions as imagined and mobilized in policies and programs. Rather than the development of more universalizing best practices, the book shows how accumulated knowledges based on the specifics of everyday built environments such as yards inform and shape the possibilities for new urban worlds that might emerge – or enable us to better recognize those already in practice.

What (Else) Might a Yard Do?

The book reveals the significance and potential of ongoing and ordinary relationships with environments, including aspects of environmental care, labours of maintenance and cultivation, negotiating difference, and emotions such as joy and sorrow. Throughout the book, I aim to provide clues into world-making processes that so often go overlooked. What is affirmed in these yards, and what is denied? What is seen, and how, by whom, and for what? Beyond environmental concerns, *Living with Yards* aims to shift conversations about how scholars and practitioners in urban studies and geography know cities, by centring the ways people live with and experience urban surroundings in these ways of knowing.

If planners, designers, and scholars are to better understand urban environments in their fullness – through all the uneven power relations and mutually entwined social and spatial processes – we must draw from familiar terrains such as everyday yards all that we can. It will be necessary to add to reductive techno-scientific models and universalizing best practices. What if a rejuvenated, amended sense of environmentalism informed planning, decision making about, and experiencing urban environments? What if that environmentalism drew more intentionally on everyday experiences of both the continuities of urban habits and the long-standing relations that shape these continuities – as well as the creative potential of already existing ruptures, understandings, and shifts in such habits? *Living with Yards* aims to see this interplay between sedimented urban habitats and habits through the creative possibilities of everyday practice and experience. Together, the chapters of the book point toward opportunities we might make to engage everyday desires for the care and cultivation of others through the habits and rhythms of inhabitation. The making and maintaining of these relations constitute emerging terrains not necessarily bounded by private property ownership drawn across the land, but rather ones embodying the future worlds of care, affect, and rhythm.

1

BUILDING
URBAN
HABITATS

The imagination is hampered in its flight.

HENRI LEFEBVRE, *The Urban Revolution*

I stand on the front porch at the door, and John tells me to walk around the side of his house – he'll put on his shoes and meet me at the back door. He points around to the north side of the house and says, "Check it out – you'll start to see the yard." As I round the corner into a large and sunny side yard mostly obscured along the front sidewalk by large bushes as tall as me, I see a vast pile of plastic gallon jugs haphazardly clustered around an open plastic garbage can and a few buckets. Some are filled with water, some empty. Roof gutters empty into a downspout, which in turn empties into a large dishpan. John tells me when he meets me outside, "This is my watering system. The garbage can is a rain barrel, of sorts. Then I fill the jugs when

it rains. I use those to water everything. It's my water conservation system, see." Although he sees these practices as quite separate from the broader urban context in which his yard is situated, municipal governance, in the form of historical decision-making about development and the building of infrastructure such as streets and sewers, and present-day regulations to do with yard and home maintenance and land use also shape John's yard experiences. In all aspects of his relationship to his yard and home, John is driven by a desire to build and create on his own terms, with the materials and expertise at hand. His idiosyncratic "water conservation system" shows this self-reliance.

John is by no means alone in placing importance on water use, water conservation, and stormwater capture. In the past three decades, in large and small cities across the United States, municipal policymakers and planners have increasingly framed similar concerns about physical and natural environments by using the concept of urban sustainability. Though details vary, urban sustainability is most often mobilized in formal governance through sustainability plans focusing on quantifiable metrics – sustainability indicators. Goals are set for these indicators, and best practices are identified and encouraged through a multitude of projects and programs. These plans and metrics frequently overlap and dovetail with ongoing citizen- and nonprofit-led environmental projects, as well as the creative making of everyday life. In short, these manifestations of contemporary sustainable urbanism have taken a firm hold in how planners, architects, and urbanists think and in the ways they know the city. Yet these plans are not always effective or substantive, even by their own measure.

Official practices influence John's own everyday practice, by shaping the context of what is legally allowable. In this way, the forces that shape city policies also shape John's habitat. In turn, John makes this habitat meaningful in the way he manages rainwater, using it to cultivate plants in other areas of his yard, and redirecting water away from the municipal sewer and treatment infrastructure.

The central endeavour of any urbanism is to dream and plan the city yet to come, while at the same time managing the city now. This effort tends to entail techno-practices that claim to know the city, and the most recent iterations of these techno-practices underlie the notion of the sustainable and resilient city. Planners and designers offer *habitats* – sometimes radically visionary habitats – that depend on quantifiable metrics and the apparatus of city codes that regulate urban form and maintain regimes of property ownership. Meanwhile, people's practices of *inhabitation* – the ways urban inhabitants actively make, do, and feel their urban environments, as well as endeavour to make a meaningful common life – often are reduced to best practices or simply pushed to the margins of sustainable visions.

Ideas about the sustainable, resilient city tend to reduce yards and other environments to their constituent physical components; in so doing, they limit capacities to support and imagine these spaces as they might otherwise be. The contemporary shifting landscape of the purpose and practice of gardening raises questions about how we might imagine and shape more sustainable urban environments differently.[1] As the techno-practices of city codes transpose inhabitants' doings into their material effects, inhabitation itself is largely written out of official imaginaries of future green urbanisms. At the same time, capacities for inhabitants to imagine city codes and aspirations are situated within specific geographies of enforcement, regulation, and policing, which residents understand as happening unevenly across neighbourhoods and the city at large. Thus, official visions are only one part of a complex terrain of green urbanism, involving much broader practices and projects than are usually considered in either official plans or critiques that respond primarily to those plans.

In particular, urban gardening, which forms the locus of this chapter, has become a central character in sustainable urbanism. In the past decade, there has been an explosion of interest in urban gardening. The role and potential of gardening within the city are presently seen as integral to making local and healthy foods more accessible, mitigating negative impacts of urbanization on habitats and water quality, lowering carbon emissions, and contributing to emerging forms of social justice, community, and sociality in urban space. A range of dynamic negotiations takes place around yard governance and activism focused on yards. Yards can be productively seen as a diverse field of motivations, practices, and understandings of city life, and that observation applies equally to the other ways of managing yard spaces that this book describes.

In Minneapolis, urban gardening has recently been the focus of major interest on the part of city government, as well as the subject of many nonprofit and activist groups stretching back at least six decades. In this chapter, I examine two recent cases: first, the adoption of an Urban Agriculture Policy by the City of Minneapolis, with associated amendments to city codes to do with allowable gardening for food production in a variety of city spaces; second, ongoing efforts on the part of a prominent nonprofit organization to promote the design and installation of residential rain gardens. Both of these efforts are quantified as part of formal sustainability indicators tracked by the City of Minneapolis, but the official vision of how urban gardening might fit into broader notions of a sustainable city remains limited.

Individuals and neighbourhoods understand and imagine *habitat* quite differently from the way city governance, designers, and planners see it. As I show in detail in subsequent chapters, the ways of knowing everyday urban environments through inhabitants' practices and experiences entail highly intimate, creative, complex, and often communal capacities. These shape, and are shaped by, residential yards. Municipal environmentalisms, such as those organized around the concept of sustainability, supplement this everyday making and remaking by governing through existing codes and incentives (themselves originally shaped by a complex mix of best intentions and diverse interests). In the process, sustainability plans and projects render urban environments without a strong sense of the different modulations of inhabitation – the very ways of knowing the city that are most meaningful in everyday life.[2]

City as Habitat

How do we know the city? More precisely, how do we know the city in relation to environment? Henri Lefebvre identifies a chasm between the expert knowledges and skills of what he calls "architects and urbanists" (planners, policymakers, designers) and the concrete practices of people's inhabitation.[3] His distinction between *the city as habitat* known through technical and skilled knowledge, and the urban as produced through *everyday lived experience*, or *inhabitation*, helps identify the dissonance underlying sustainability policies and projects. By focusing primarily on measurement and best practices, sustainability efforts on the part of cities such as Minneapolis render urban environments in terms that reinforce the city habitat as static and inert. However, official city programs are not the only means through which sustainable urbanism is deployed. Examining the many concrete forms of sustainability reveals a diversity of perspectives not encompassed by Lefebvre's habitats void of meaningful inhabitation.

In the most general terms, *habitat* usually means an organism's or species' surroundings, implying the kinds of food, space, and resources it may require and prefer. Lefebvre finds this functional view of habitat distressing for the way it so easily makes possible simplification and reduction of the diversity of urban life, and for how it has come to subsume the practices of inhabitation. The very notion of what it means to be human becomes reduced here – even beyond, or perhaps beneath, what it means to be animal – in what Lefebvre describes as "a handful of basic acts: eating, sleeping, and reproducing. These elementary functional acts can't even be said to be animal. Animality is much more complex in its spontaneity."[4]

This habitat view overwrites the importance of concrete urban life by repressing the difference within modulations in how people live. Lefebvre writes, "Habitat was imposed from above as the application of a homogeneous global and quantitative space, a requirement that 'lived experience' allow itself to be enclosed in boxes, cages, or 'dwelling machines.'"[5] He finds these reductive moves central to the devaluation of the creativity and potential of everyday life, and he uses the term translated as *habiting* to convey how he sees everyday life as full of dynamic practices. For Lefebvre, the most resonant notion of dwelling was a place in which the creative poiesis of everyday life unfolds with meaning and potential in surroundings that afford active and continual making. In contrast, habitats offer little potential for appropriation or for "making them one's own," and so, in comparison to dwelling, are an impoverished effect of a view of the city that obscures everyday life.

Habitats, and inhabitation, are never universally equal in their conception, execution, or experience. Black and Indigenous studies, subaltern studies, queer and feminist scholarship, among others, have clearly documented the uneven impacts and afterlives of even the most well-intentioned urban projects. These sometimes violent structural oppressions differ in impacts and experience along axes such as race, gender, ethnicity, sexuality, and dis/ability. The state and private interests of capitalist markets know these environments through apparatuses of measurement, administration, and bureaucracy. Circulating in and through such worlds is the political potential of what cannot be captured or reduced to these legible concepts and relations – the modest domains, practices, and everyday life in excess of functionalist categories. This is *the possible* within the ways people live, and it cannot be contained by habitats. Here, there are possibilities for socio-spatial transformation within the concrete ways we already live collectively and communally, even as dominant economic, juridical, and political systems impose and capitalize on our individuated nature. In this sense, the habitat-inhabitation relation is the *ongoingness* of making from above (urban planning) and making from below (everyday micro-geographies and neighbourhood practices). Thus, habitats can never be understood only on their own terms, but must be approached at multiple scales, and in terms of how people live in, through, and against habitats.

Several approaches from urban planning and design literatures have emphasized this way of knowing habitats. In 1986, an innovative book was published about housing and everyday life. The provocative title, adapted many times over in planning literature, conveys the central argument and approach: *Housing as If People Mattered*.[6] Marcus and Sarkissian identify a major dissonance in the ways planning and design projects are conceived, designed, developed, and managed: the lack of a meaningful presence of inhabitants and their experiences in design and subsequent management processes. The authors drew attention to the fine-grained ways people lived *in* and *with* space – for example, the ways people made entries their own with decorations under varying spatial conditions, or especially the many ways children actually played (or not) in common outdoor spaces designed specifically for them (or not). The distance between inhabitant and designer so beautifully and respectfully rendered by Marcus and Sarkissian is important far beyond their specific cases of public or government-designed medium-density family housing in 1970s and 1980s. It speaks to the same general dissonance that reverberates throughout modern urban planning and broader urbanisms in the past 150 years.

Amongst the forest of urbanisms informing contemporary urban studies, *everyday urbanism* – one recent approach at the intersection of urban design, architecture, and landscape architecture – has drawn in particular on the potentiality within the concrete practices of everyday life.[7] Proponents argue that it is through engaged and nimble design practices, including attentive and respectful observation of what is already happening in a place, as well as involvement and design *with* the people who live in and through proposed designs, that meaningful urban places will emerge. The initial efforts in the mid-1990s toward defining everyday urbanism were met largely with silence and dismissal from increasingly popular, and now thoroughly mainstream, design thinkers like those of New Urbanism. That approach tends to favour form-based solutions that rely on particular productions of nature and nostalgia, as well as exclusionary senses of place, in order to be legible.[8]

More Sustainable Habitats?

"Just start measuring things! It doesn't really matter what they are – don't worry too much about which indicators to use. Just get a baseline measurement, and then don't be discouraged if things don't improve all that quickly." So said the City of Minneapolis sustainability coordinator at a symposium on urban ecologies for scholars and practitioners.[9] This advice captures much of the urgency and immediate jump to measurement that has reverberated throughout applied planning literature on sustainability and green urbanisms for the past three decades. Sustainability policies and projects present a recent iteration of the dissonance Marcus and Sarkissian identified between design and planning, on the one hand, and inhabitants' experience, on the other. They also echo Lefebvre's theory about the ongoing dialectic of modern cities known and planned as *habitats*, and the urban as *lived* by inhabitants in daily life.

Planners have tried to meld "green design" principles often developed in terms of single buildings and green architecture, with socio-spatial scales such as neighbourhood, city, and region. This has taken a variety of forms. Throughout all of these moves, similar pressures to quantify, simplify, and reduce the world to knowable systems shape the possibilities planners and designers imagine for cities and neighbourhoods yet to come. For municipalities buffeted by pressures from economic crisis and neoliberal austerity, environmental uncertainty and disaster, and persistent socio-economic disparities, sustainable urbanism presents another layer of possibility, and also constraint, that takes shape in relation to existing institutional structures and pressures.

Urban sustainability as a planning and policy concept first emerged in the United States in the early 1990s in cities such as Seattle, Santa Monica, and San Francisco.[10] These cities especially drew on the "three E's" – environment, economy, and equity – first laid out in the UN Brundtland Report in 1987, as a means to capture the multi-faceted nature of sustainability as a guiding concept.[11] Cities in the United States and beyond continue to use this formula as a basis for sustainability efforts. However, institutional arrangements for incorporating sustainability within existing urban governance vary across time and place. Separate departments or offices were often introduced in some of the earliest cities to adopt sustainability policies, such as San Francisco. In cities that adopted such policies a bit later, including the city of Minneapolis, sustainability goals were integrated across city departments, often overseen by a coordinator and small staff.

The city has become the central scale around which sustainability and resilience policies are enacted. Such policies have become another mode through which to neoliberalize local entrepreneurialism, with cities (and sometimes regions) competing for capital investment through the promotion of green projects of different types. In the United States, urban-scale sustainability imagined through calls for acting locally has meshed seamlessly with much longer traditions of resistance against European models of centralized national and regional planning.[12]

In the examples discussed below, formal and informal approaches to imagining and enacting more sustainable urban habitats reveal a varied and power-laden terrain organized around multiple understandings of urban gardening. This can be seen in the first case below, as the City of Minneapolis has tried to overhaul building and zoning codes to allow more urban agriculture. In the second case, focused on rain-garden advocacy and urban gardening, I discuss the ways in which one nonprofit devoted to urban gardening in the form of rain gardens both reinforces and disrupts this city habitat perspective. Finally, the third case study focuses on how the cultivation of trees provides further layers of everyday engagements, experiences with multiple temporalities, and concerns about urban environments spanning property bounds and straightforward public/private distinctions.

Urban Sustainability: Restrictions and Aspirations

In Minneapolis, municipal sustainability efforts echo the above trends, with efforts primarily focused on developing and tracking sustainability metrics. The city council formally adopted sustainability as a key principle in 2003, followed by a series of actions understood by the city to build on past environmentally oriented policies and projects.[13] These included developing a program to identify and track key sustainability indicators and ten-year targets for those indicators, as well as amending ordinances such as the city's zoning code to be consistent with the sustainability plan. In the original 2003 resolution, *sustainability* is deployed as an umbrella concept intended to synthesize and guide environmental decision-making in a more coherent way, and eventually "integrate the Three E's, Environment, Economy and Equity (including social justice); coordinate efforts; garner buy-in; and increase the effectiveness of our ongoing programs and investments."[14] A small office was established that currently houses several staff, including a sustainability coordinator.

Sustainability indicators dominated these efforts in Minneapolis until recently. In 2005, the city held a series of public roundtable discussions to inform the selection of approximately twenty-four indicators, and a focus on sustainable growth was incorporated into the city's comprehensive plan in 2009. For several years, these sustainability indicators were modified periodically and roughly grouped around the initial concepts of health, environment, and social equity. As the indicators became further stretched across these areas at an increasingly granular scale, the City of Minneapolis pulled back on indicators and now incorporates assessment into "Results Minneapolis," a series of reports about long-term goals across diverse issues.

Environmental Possibilities in the City of Codes

Codes work as a dynamic regulatory apparatus in contact with our most intimate lives, stretching into seemingly quite private spaces such as dwellings and yards. The priorities set out by city and regional governments through municipal codes have a profound impact on the making of city habitats. Municipal codes operate in two primary registers at once: restrictions enforced through regulations, and encouragement in the form of incentives. Municipal codes such as building and maintenance codes play an important role within this duality by defining in highly detailed and spatial terms what is allowable or possible. As one experienced policy aide to a city council member explained to me:

> In Minneapolis, we aren't at the point of saying you *can't* have lawns, but we have decided it's *allowable* not to have a lawn. It's possible. At the top levels, there hasn't been a shift towards really saying, "This thing that you've been doing forever [cultivating a grassy lawn] is now no longer acceptable. You can't do this anymore" – even though that would actually make a significant difference towards stated sustainability goals for the city.

This description points to the conundrum between managing the present city and planning for the city yet to come.

Interest in more sustainable and resilient urban futures has helped drive renewed interest in codes, best practices, and policy transfer between places. For example, urban agriculture has multiple web-based portals through which different actors such as government policymakers, politicians, and advocates compare notes and share specific policy language about urban livestock, community and rooftop gardens, farmers' markets, and the marketing and selling of homegrown produce. In Minneapolis, the sustainability coordinator participates in a national network of such coordinators by attending periodic conferences as well as by participating regularly in a closed web-based forum to compare notes about local policies and best practices.

The first regulations of private property enacted in the 1920s were generally relatively minimal documents restricting land uses in certain areas. Regulations rapidly proliferated with the development and building boom in the postwar years, as a means to protect and maintain property values. These early zoning and building codes served to further segregate urban and increasingly suburban areas – thereby concentrating poor, immigrant, and nonwhite populations in squalid living conditions and limiting their access to new suburbs. The increasingly systematized and "scientific" view of urban environments through large-scale urban projects such as the interstate highway system rendered cities as composed of discrete and knowable systems. Codes likewise became significantly more elaborate. For example, by 1963, the two-page Minneapolis city code from the 1920s had exploded into a large comprehensive code of more than a thousand pages, very similar to the city code document of today. In the process, activities previously not regulated at all – such as urban agriculture and animal husbandry – were dramatically restricted, with allowable uses and physical spaces spelled out in the code. Now such municipal-scale regulations are being reimagined in terms that resonate with neoliberal aims to roll back the state: flexibility, agility, and removing barriers to innovation and enterprise. In Minneapolis, city codes serve as a key lens through which to see how sustainability governance and planning shape how planners and policymakers see the possible.

Habitats for Minneapolis Urban Agriculture

In 2008, Mayor R.T. Rybak, in conjunction with the Minneapolis Department of Health and Family Support and the Minneapolis Sustainability Office, embarked on an effort named "Homegrown Minneapolis" to rethink the role of the city in supporting and shaping local food systems. Dovetailing with sustainability indicators, the initiative studied and compiled recommendations for policies to create a "healthy, local food system."[15] The Homegrown Minneapolis report (2009) required the city to put together the first "Urban Agricultural Policy Plan." Committees were assembled to translate the aims of the plan into zoning code amendments. Two years later, the city council adopted the Urban Agriculture Policy Plan. The associated text amendments defining, clarifying, and regulating how food might be grown and sold within the city were adopted in 2012. Historically, growing food was largely written out of legal and regulatory understandings of urban land use when the first major comprehensive code was assembled in 1963. The key function of the current Urban Ag Plan was to expand allowable food production, processing, and commercial exchange in the city through amendments to city codes. This process largely relaxed relevant regulations governing urban space, but also involved identifying and defining urban agriculture practices and associated material requirements in terms of land uses, gardening structures, and activities. For the current Urban Agriculture Plan, discussions around adding new definitions and amending existing code took the better part of one year, preoccupying community gardening enthusiasts, food justice advocates, and organic and local food activists in the area. Throughout, people involved with various types of gardens were integral to this process, yet the metrics mobilized overlooked important aspects of their experiences.

The urban agriculture amendments to city codes reinforced some existing understandings of urban environments and challenged others. Receiving much of the attention and press, these definitions now include several types of large-scale food production, including *market gardens* ("an establishment where food or ornamental crops are grown on the ground, on a rooftop, or inside a building, to be sold or donated") and *urban farms* ("an establishment where food or ornamental crops are grown or processed to be sold or donated that includes, but is not limited to, outdoor growing operations, indoor growing operations, vertical farms, aquaculture, aquaponics, hydroponics, and rooftop farms").[16] For both *market gardens* and *urban farms*, the ability to sell fresh produce at the site where food is grown (on a limited number of days per year) was a major change. The definition of these new land uses was celebrated and widely promoted by the city and gardening advocates as allowing increased opportunities for economic development and facilitating entrepreneurial drive across diverse populations in the form of for-profit food production in the city. Additionally, the definitions recognize for the first time that food production in the city may take a variety of physical forms – including commercial-scale aquaponics, hydroponics, and living-roof systems.

Although smaller in scale and less frequently touted in press releases, growing food in residential yards garnered much debate, which points to the wide variety of urban gardening scales, practices, and understandings on the part of urban agriculture advocates and municipal planners. Discussion and debate about small-scale gardening structures such as arbours, raised beds, cold frames, and hoop houses were framed around multiple understandings of the purpose and use of residential yards.[17] Regulation of yard spaces generally is located within one of two main municipal purviews/departments: planning (land use and zoning), and housing maintenance (outdoor upkeep such as overgrown lawns and broken windows). In both cases, outdoor space is defined as front, side, and backyards – with different allowable uses and maintenance guidelines for each. Front yards are significantly more limited in the range of allowable uses, and backyards are much less regulated. The planning department's purview with respect to yards is complaint driven, and the enforcement of maintenance regulations is a combination of complaint-driven enforcement, and annual visual surveys done by the city.

Debates about the proposed urban agriculture policy changes were constituted largely from two main perspectives. First, urban agriculture proponents tended to focus narrowly on the capacity of a single yard to produce food based on the best growing conditions (usually full sun), regardless of regulatory distinctions between front, side, and backyards. In contrast, city planners had broader conceptions of yard uses, especially in front yards, and they expressed concerned that the preservation of view corridors and neighbourhood character was potentially threatened by increased productive food gardening. These arguments against expanded allowable gardening structures such as arbours or raised beds included concerns and protests about "unsightly" disruption of the collective environment along the front of city blocks. Urban agriculture advocates countered this resistance to expanding allowable structures and uses of front yards with very different visions of urban neighbourhoods. In advocates' views, agriculturally productive land could exist in every yard, with a range of structural supports to maximize food production. In the end, the revised yard regulations do expand allowable urban agriculture possibilities, especially in front yards, with some compromise. Because the municipal planning and maintenance departments are involved primarily when neighbours complain, it remains to be seen how the revised urban agriculture policy may shift yard practices and norms, and how this may vary across the city.

Official codes and ordinances telescope from the broadest category of land use down to the most minute, including, for example, legal regulations with respect to dog feces within the city. In one yard, prominent self-made signage reminded passers-by of such regulations.

Codes also serve as a point of negotiation around which different perspectives pivot, due to the fact that they literally define urban life (or at least purport to) in spatial and material terms. Clearly, urban agricultural code amendments are negotiated terrain rather than a series of black-and-white decisions, and what constitutes a best gardening practice is not always indisputable, even if environmental activists claim otherwise. A tension between *gardening as food production* and *gardening as beautification* has emerged in these debates around city codes in Minneapolis. Planners working for the city felt the need to push back against urban agriculture activists who primarily understood yards in terms of their biophysical and productive capacities. The planners with whom I spoke considered this aspect of their job as taking the collective long view. They understood their role as looking out for unforeseen negative consequences of a total shift (however unlikely) towards urban food production in residential yards. They also considered it their responsibility to keep a more diverse range of uses and meanings in play for residential yards, even if this meant limiting some of the allowable urban agriculture uses.

The process of revamping codes in place since 1963 raises questions about how urban agriculture and urban gardening are defined at very fine scales. This process reveals what is made legible in the policy context – how these definitions emerge from, and also produce, a narrow rendering of urban gardening. Urban agriculture activists saw little value in modes of gardening beyond production, belying the fact that gardening embodies different meanings, life experiences, and collective capacities depending on where, how, and by whom it's practised. Taken together with sustainability priorities from the city, the debates and code revisions show how simple metrics cannot fully embody the socio-spatial dimensions of more sustainable urban habitats.

Aspirational Sustainability through Rain Gardens

In addition to food-producing gardens, the City of Minneapolis tallies the number of rain gardens each year as one sustainability indicator. Rain gardens usually take the form of a shallow depression planted with a variety of plants, and are designed to capture rain water from downspouts or impervious surfaces such as driveways and sidewalks. Ideally, these plants slow and divert stormwater runoff, keeping it from polluting storm sewer systems. Because the typical residential roof in Minneapolis uses asphalt tiles, the city advises that rain gardens generally not be food producing, due to the potential for chemical residues in the water to be taken up by edible plants. Usually native plants are recommended, as they often tolerate local climates better without as much need for added water or chemicals. A particular urban environment is imagined through rain gardens – one that involves a reclamation of territory from "exotic" species such as turfgrasses, and that redirects the movement of water and pollutants vertically near the source, rather than horizontally across city landscapes.

In Minneapolis, one nonprofit organization, Metro Blooms, has been central to efforts to improve water quality through rain gardens. The organization grew with initial support from the city and local activists in the 1960s, and the focus has shifted from an emphasis on beautification toward a direct engagement with water-quality issues and stronger environmental advocacy in the past decade. Metro Blooms developed a series of workshops in 2005 to educate the public about and promote rain gardens in particular, as one type of urban gardening that can contribute to better water quality. Metro Blooms now partners with a variety of government agencies and neighbourhood groups. Its workshops are promoted throughout the Minneapolis metropolitan area, including website links and promotion by the city of Minneapolis Sustainability Office and Public Works Department. Because these gardens are often misunderstood as "messy," "shaggy," or unattractive by neighbours or passers-by, an educational mission is often built into rain-garden design. For instance, Metro Blooms gardens often include signage reading "I am a raingarden. I capture rainwater to protect our water resources." Signage and literature on rain gardens assure homeowners that these gardens can be beautiful and can trap mosquitoes rather than encourage them, and that the gardens require little maintenance after plants become established. This type of information points to the work necessary to address many residents' anxieties and their reluctance to adopt this different form of gardening.

One of Metro Blooms' ongoing initiatives, Neighborhood of Rain Gardens, scales up from a focus on individual residential properties to conceiving and

encouraging rain gardens at the scale of neighbourhood. This has taken different forms in different neighbourhoods, often determined by partnerships between Metro Blooms, funding agencies, and neighbourhood organizations. Two recent Neighborhood of Rain Gardens projects of differing scales show an evolving approach on the part of Metro Blooms regarding how neighbourhood rain-garden initiatives are communicated to residents, how individual rain gardens are designed and installed, and the role of measurable outcomes. The first was a large-scale project of more than 120 rain gardens, designed and installed in 2010. While the number of gardens and associated storm water metrics reached the targets, the designers found resident involvement varied widely at all stages of the project – initial interest, design and installation, and ability and willingness to maintain gardens over time. Property ownership status may have affected participation in the program. Metro Blooms designers felt that renters – who made up 53 per cent of the households in the targeted neighbourhood – may not have had as much involvement in outdoor spaces in general and therefore may

not have felt as empowered or capable of maintaining the gardens.[18] As one landscape designer elaborated, "We knew a lot about the watershed, but not as much about the people-shed." Still, the project has made considerable difference in the appearance of the neighbourhood's yard spaces and the way the residents are experiencing them. One resident described it this way, "Even without everyone taking care of their gardens, you see a different way of having a garden." Metro Blooms now focuses volunteer days around seasonal maintenance, such as clearing mulch and debris from rain gardens and street drains in the spring.

In a different part of the city, a smaller Neighborhood of Rain Gardens project was undertaken, with a decidedly different approach to involvement on the part of residents. Unlike earlier projects, where volunteers dug up and installed gardens, residents were engaged early on to work with designers and to collaborate to help one another prepare and plant the gardens. This focus on residents' direct involvement in planting gardens and on neighbourly help was intended to foster feelings of ownership over their gardens and to "build community." The goal is that, over time, others in the neighbourhood will see how their neighbours were able to do their own rain gardens, and will be inspired to consider the idea themselves. Measurable impacts such as quantity of runoff captured by gardens of this neighbourhood project were more modest than in the larger-scale project, due to a smaller number of rain gardens. However, Metro Blooms designers felt that the different approach led participants to master the techniques first-hand, perhaps feel more confident about their own rain gardens, and establish an informal support network with neighbours.

From an environmental advocacy perspective, the recognition of relationships between neighbours is increasingly important. "Community" is imagined as being "strengthened" and "built" through the encounters between people and gardens, and this idea has become a central part of Metro Blooms' discourses around their programs. Participants in the small rain-garden project described above self-consciously joked at the first installation demo as they chatted and drank coffee on a chilly early morning one weekend: "Well, we're getting off to a late start – but we're building community. That's what it's about, right?"

One year later in this smaller project, some of the rain gardens have languished and not received the care that would help them flourish. By the city's sustainability metrics, they count as rain gardens, but they might not be reaching the goals laid out by Metro Blooms and the neighbourhood organization. The city's number of rain gardens sustainability metric doesn't quite capture the social connections surrounding rain gardens, or advocacy groups' work to influence changing urban gardening practices.

The difference in these cases between official city codes and expert advocacy may not matter much if both deploy reliance on narrow slices of specialized knowledge about urban environments. However, several key differences emerge in the latest iterations of the Neighborhood of Rain Gardens program in Northeast Minneapolis and the city's official plans and metrics. The rain-garden program, while still limited in scale and relevance for many people's yards, endeavours to offer the kinds of socialities through which meaningful environmental practices such as rainwater capture in individual yards could take hold and become part of broader everyday life for urban inhabitants. Through the mobilization and fostering of affective encounters in workshops and informal gatherings, as well as in its thinking in terms of multiple temporalities in conjunction with plants, people, and neighbourhoods, Metro Blooms provides additional dimensions to urban environmentalism beyond city habitat perspectives focused solely on measurement. Still, the rain-garden rhetoric revolves around best practices such as how to install the gardens through very prescriptive steps, as well as the kinds of plants suitable for them.

This rain-garden advocacy is based on fairly narrow views of an archetypal homeowner who lives in a single-family (often suburban) house.[19] So, while taking this kind of project into account expands the usual critiques and analyses of sustainability policies beyond planning and measurement, significant limitations remain to moving toward a more robust and more meaningful urban environmentalism.

Rain Gardens 2.0: Adapting to Rain Garden Affects and Rhythms

Metro Blooms tried out a new tactic for inviting a broader range of people as the Neighborhood of Rain Gardens program developed into its second year in Northeast Minneapolis, with the continued involvement

of individuals in the area and from the neighbourhood group. This tactic was an informal information session held in the backyard of a particularly keen couple who had installed several rain gardens with the help of the program in its first year, and then a short walking tour of rain gardens in early summer. Some participants from the first round of rain gardens came to tell people about their experiences, and funding from Metro Blooms and the neighbourhood organization helped provide snacks and a bit of beer and wine. People who heard about the program came to find out more. Some people brought their neighbours along. After a loose presentation by one of the Metro Blooms designers about rain-garden basics, projected onto a portable screen as chickens clucked underfoot, the group meandered toward several rain gardens built and installed the previous year. Residents who had rain gardens told the group about their experiences, and pairs and trios of people chatted and compared gardening notes as they walked from garden to garden.

The lived experience of the evening, with all of the mingling – people making time to meet one another on a busy weekday evening, ask others how they'd heard about the event and what they hoped to do in their garden this season – as well as the relaxed approach to conveying the technical information about rain gardens, highlighted the sense of social connection through rain gardens, which organizers had hoped for. It also served to make rain gardens seem manageable, desirable, and important, through the narratives of residents and neighbours. The structure of the event also allowed for individual stories and perspectives to come through, even when not totally in line with the prescriptions for rain-garden design and installation. For example, one participant championed the program in general although he hadn't participated. He told me he hoped the neighbour he brought along, who is a new homeowner and a bit overwhelmed by what to do in his yard, would seriously consider the rain garden program as a way to become more engaged with his yard – to, in essence, take some "baby steps" toward more involved gardening. The approach of Metro Blooms and the neighbourhood group points toward more relevant and meaningful registers through which urban environmentalisms might take hold. While still reinforcing dominant conditions and status quo understandings, this approach does open up relations through which people may be affected – by neighbours, plants, soil, and water.

Loss, Growth, and Care in the Sustainable City

If the annual food gardens and multi-year rain gardens embody some of the key complexities of urban gardening, trees in urban forests reveal long-term relations of growth and care on the scale of human lifetimes and beyond. Across the Twin Cities region, interlinked canopies shade swaths of more affluent neighborhoods, and sparser plantings of mature trees dot less affluent areas. Urban forests intersect with sustainability goals in several key ways such as providing shade for buildings and streets, thereby reducing cooling costs during hot months; and stabilizing soil and preventing runoff. Trees also significantly shape the character of streetscapes, something important to inhabitants and city planners alike. As studies have shown, the presence of trees can often be connected directly with socio-economic status indicators such as income, wealth, and property values, and can provide insights into urban environmental change over time.[20] Experiences detailed below and in the next chapter underline deep-rooted relations between people and trees, and highlight a dimension of urban environments yet to be fully appreciated and integrated into formal initiatives. The case study of a destructive tornado event and related damage to neighbourhood tree canopies reveal underlying temporalities of cultivation and care over long periods, which are often not the focus of sustainability metrics from municipal perspectives.

An Urban Tornado and the Loss of Mature Trees

Sometimes a sudden event or disruption reveals a relation otherwise obscured by everyday life. Such is the case with respect to a massive tornado that travelled through parts of the North Minneapolis study area the year before I undertook my fieldwork. A year later, Betty and I sit at her dining room table on a bright sunny weekday morning. It's not long into my first yard visit with her, and, after responding to some of the particulars of the structured questions, she has started telling me about the tornado. In the late afternoon of Saturday, 21 May 2011, a tornado travelled right along her street, just one stretch of the path the tornado made through urban and suburban neighbourhoods.

For Betty, the storm has been transformative. Both Betty and her husband of more than forty years, Sid, are retired educators – Sid a longtime principal, and Betty a high school art teacher. The past five or so years have been full of transitions for them. Betty was in an accident and suffered a brain injury, which put her, as she described it, "out of commission" for two years. Sid has had his own medical issues, and, when the tornado arrived, they were preparing for the trip they made the following day to the Mayo Clinic in Rochester, Minnesota, in order for Sid to have surgery, for which he'd been waiting a long time. In the midst of these preparations, a tree was blown right through the roof and a bedroom of their house. The damage to their home was extensive, and they were relocated to an apartment building in downtown Minneapolis for nine months while the house was repaired and insurance claims sorted. Betty has since had time to reflect on the experience, and she talked to me about how one grows out of such a challenge in unexpected ways.

This growth has crystallized for Betty in her feelings about the drastic loss of mature trees along her street and in her neighbourhood. She tells me,

"When I got back, and the more I had time to sit down, outside – outside, too, I had to be outside – I realized that sometime things look pretty bad but they can be good." Her voice is full and wavers a little with emotion. "And the storm was really bad, but I felt like I grew. In places I never would have – so I'm appreciative of that. Cause I never even gave it a thought about the trees on the block!" Betty began to laugh as she mimicked her earlier self. "'Who cares!' I'd say. 'Maybe we should cut that one *down*, so I can *see* better!' Then I realized what trees are for."

Trees are revealed as a measure of time through their sudden destruction and absence. While Betty mentions the cooling shade they had provided in a hot summer, the trees also become a measure against which Betty understands her own lifetime. "My biggest sadness that brings tears, is when I think about all the trees that are gone, that were big and mature, on our block. And I will never be around to see that again. Cause thirty or forty years" – she smiles a soft smile at me and her voice quiets – "that would be nice, but I won't be here." This awareness of her own lifetime leads to impatience with the replacement trees the city has planted along the boulevard. Betty has a new and fervent interest in these trees. "It's so weird how much I cared about the trees [after they were first planted]." Despite careful watering and attention, the first replacement tree didn't make it and had itself to be replaced. "Of all people! *My* tree would be the one to die!" She got to know city employees, who answered her questions and checked up on the tree, and even sent her a note of thanks. "I check that tree and water it, and I know how important that tree is gonna be to me." And she has tried to convince neighbours who complained to her about paying for the water that they will benefit from the trees in the long run.

Walking through her yard with Betty reveals additional dimensions of her experiences there. She brightened, and in her voice there was a shift in feeling from somewhat flat reporting to inhabiting memories. The yard for Betty clearly holds a lot of experiences that have passed. The recent physical limitations of both herself and her husband have meant that her attitude toward the yard is one of utilitarian maintenance with which she feels they can barely keep up. She is tentative in discussing her garden engagements, revealing a distance and hesitation as she talked about the plantings and spaces. Still, the storm seems to have shifted this sense of obligation toward one of care, which is directed at the new plants she has planted as well as the boulevard tree that is to be replaced by the city. Betty's experience of the sudden loss of trees reveals a sudden sense of their long lifetimes and highlights transitions already underway for Betty and Sid with respect to their shifting capacities and engagements with their yard.

Sustainability beyond Habitat?

Specialized sustainability perspectives render urban environments as primarily physical habitats through city codes that regulate allowable uses. As nonprofit advocacy projects have attempted to reach beyond this narrow focus in an effort to incorporate concerns about community, participation, and changing norms about urban gardening, a more expansive view has emerged. This view starts to include how people interact not just with physical space, but with one another. Advocates and designers understand that such expanded views are necessary to further the success of rain gardens. In turn, it is understood these rain gardens have implications for water quality across watersheds, ultimately contributing to official stated sustainability goals.

The urban agriculture, rain-garden, and tree-recovery efforts discussed above disturb dominant understandings about what constitutes urban gardening and a garden – especially a garden located within a residential front or backyard. The multiple discourses discussed above embody different imaginaries about how sustainability is understood, and how changing practices may reshape and contribute to a more sustainable urban life. In the case of developing urban agriculture policies, the process of working out the definitions and details of city code amendments reveals sometimes conflicting understandings about how and where urban agriculture should take place. Urban agriculture advocates argue that urban land such as yards must be understood in terms of available resources such as sun and water, which override concerns of planners and some residents for the character of the street or for the distinctions between front and backyards. These negotiations take physical shape in the city code apparatus, defining the allowable materials and contents of spaces for food production in the city. Rain-garden advocates and experts are shifting toward recognizing the importance of social relations in the long-term success and failure of sustainability goals. Furthermore, the spatial scale of the environmental interventions of rain gardens intersects with new and existing socialities at fine-grained scales of city block and neighbourhood. Trees, especially boulevard trees, add further layers of growth, temporality, and care right at the edges of public and private domains.

Such efforts still surely can suffer from narrow ideas about homeownership, individual acts incommensurate with the scale of environmental challenges, and neoliberalizing tendencies to shift responsibility for common goods into increasingly private domains.[21] Still, it is important not to dismiss as a whole the specialized knowledge of urban environmentalisms currently underway. Just as the socio-political richness of everyday life may be obscured from a planner's gaze, the layers of development history, apparatus of codes, and infrastructure beneath yards may be largely obscured from inhabitants' daily perspectives.

These debates and projects point to the importance of better understanding dynamics around environmental efforts beyond a narrow focus on measurements and indicators. Variation in efforts often considered part and parcel of urban sustainability becomes flattened by policies and critiques alike focused solely on metrics and technical solutions – the variation in understandings and practices may not be sufficiently captured with this kind of habitat-thinking. Indicators and associated measurements can define targets but may not provide adequate insights about how those targets might be reached in a particular place and depending on social differences. What we need is not simply sustainability plans that lay out indicators and measurements, but studies that examine the processes and practices that recognize larger visions and possibilities of more sustainable urban life, and which implicate a complex and broader terrain of people dreaming and making the city yet to come.

Yards offer a means to frame environmental issues in terms beyond habitat, even as most official and advocacy literature focuses instead on systems approaches (such as stormwater management). The next chapter examines how people inhabit yards through a range of embodied engagements – from the barest maintenance to intensive and artful cultivation.

2

RHYTHMS
OF
INHABITATION

"A mess," Michael describes his yard, laughing. "Ahh, what can I say?" He gestures to the north, towards his neighbour's yard. "It's relative. I look at it from the point of a formal garden on their end, and my yard is down here," gesturing toward the bottom end of an invisible ruler. We sit on the front steps in the shade on a hot July afternoon. "No, it's not the worst it could possibly be – I do *mow* it. But that's about all I do!" Michael is sixty-two, has been unemployed for more than ten years, and now volunteers as a freelance stage director in the theatre. He has a sharp wit. As I try to describe my project through the front screen door while handing out fliers, he has a wisecrack for everything I say. But I stick with it, and he says he might as well answer my questions, as he has nothing better to do. We both think it will be a swift conversation as we sit down on his front steps. Even I am surprised that our conversation unfolds over more than two hours. He tells me, "Now that you got me going on this whole thing, there's a lot more about plants and yards than at first blush I would have said. Ah, you're getting these odd stories out of me!" It's clear through the telling of these "odd stories," and the way he shows me around the space, that

Michael's engagement with his and nearby yards – while not immediately apparent or embodied through practices such as a lot of maintenance or active gardening – runs deep within his perspectives on attachment to the past, other organisms, and neighbourhood life.

These dimensions of his engagement are perhaps surprising because Michael's yard could not be more nondescript, especially in relation to his immediately adjacent neighbours' yard. A distinct physical boundary delineates Michael's yard from Adrienne and Dave's. No fence, but a dramatic difference in degree of cultivation and contents. To the south, Michael's front yard: a relatively empty lawn pocked with dirt patches, three foundation plantings around the front stoop, which are now full-blown trees. The space is fully shaded in the afternoon by a large ash tree on the boulevard, along with the umbrella-shaped canopy of one of the last old-time majestic elms growing on the boulevard across the street. The backyard is not very different – irregular grass, more sun, a chain-link fence. Adrienne and Dave's yard to the north: dense, lush. Sculpted and cascading fountains and waterways are punctuated with carefully placed specialty ornamental trees, shrubs, and boulders. Paths of flat stepping stones wind past small fairy figurines, tiny buildings, and other odds and ends that make miniature landscapes within a landscape. For the past decade or so, since they installed their first water feature in their backyard, Dave and Adrienne and their two teenage children have become more and more engaged with shaping and reshaping these yard spaces. Even before this, the yard, and house, had been continually made and remade over time. But in the past three years, the front yard has become more and more elaborate.

This fairy garden yard has become a beloved neighbourhood highlight and quasi-public space. People from blocks around know the yard, tell me about it, and regularly make forays past it on foot in order to experience the cool microclimate from the water features, sit on the benches placed along the sidewalk, and walk up into the front yard and even around the side and into the backyard. Michael chuckles with a twinkle in his eye in response to my question about what it's like to live next to their yard. "There are a few of us who laugh and giggle and think maybe it's gone a bit too far. But everyone in the neighborhood just loves Dave and Adrienne so much, they're just such kind and wonderful people. *Bemused* would be the right word. And I think I also say, 'Everyone should have a hobby!'" He laughs and continues, "Who am I to judge? And yet, I do feel some guilt that they put all of that time and effort in, and [pause] I just don't care – I'm sorry!"

Even the most minimal yard invites and demands some degree of maintenance and care – if nothing else, there is grass to be mown, and the city may impose a fine if it is not. The spectrum of engagement with yards spans Adrienne and Dave's highly elaborate practices of cultivation, Michael's minimal-maintenance approach, and an apparently simple inhabitation that entails being present in yards and attuned to one's senses and to others. We can understand lived experience with environments through the material engagements of inhabitation – as a spectrum of engagements that unfold as a continual process of producing space and time. These engagements become meaningful when situated in rhythms of social life, and they have political stakes as well – that is, they have implications for how we conceive both environment and the city. The people who inhabit yards – and home interiors – can directly shape and design them, and make them their own, often without undue financial strain. In my study areas, this is done largely without expert help or hired labour. So these mundane, familiar spaces can be important lenses into worlds of everyday life. They are bound up with – but never completely constrained, or explained, by – the parameters set out by expert designers, real estate developers, formal policies, private property boundaries, and social norms.

Inhabitation beyond Habitat

If habitat is only partially adequate for understanding urban life, then what else must we know? *Habitat*, as a largely reduced and quantified physical environment, can be understood in dialectical relation to *inhabitation* – all the lived experiences and practices of daily life that together shape inhabited environments. The two are always in constant formation together, each affecting capacities of the other. This ongoing-ness of environments often gets lost within planning perspectives on cities, which instead see them as largely static and quantifiable. In the previous chapter, I argued that when we see outdoor domestic space in the context of contemporary urban environmentalisms through this lens of habitat, we reduce it to something to be measured, regulated, and maintained within a framework of private property ownership and manageable nature. As this chapter, and the subsequent two chapters, will show, lived experiences with these spaces – the diverse ways people *live with* yards – involve a surplus spilling out beyond measurement.

Living with invites and demands particular configurations of care for others. These configurations might take shape in different ways, but they share the reality that care has a certain range, from imagining and building new creative relations to the more quiet work of tending, nurturing, and shoring up. Without experiences and skills, caring labours can be hard, and they might be the site for formal and informal learning. With certain experiences and skills come possibilities for new and different relations to emerge. People live with yards through embodied and skilful cultivation, keeping pace with maintenance, rhythms, and multiple experiences of time, as well as the ongoing relations of response and attunement to built environments through being and becoming with them. Together, this *living with* is constantly shaped by, and shaping, how people negotiate social relations through space and time.

Becoming a Gardener: Cultivation as a Purposeful, Creative Practice

Cultivation becomes an attunement between human bodies, plant bodies, surroundings, and rhythms over lifetimes. The most accomplished gardeners whom I met, even with the relative complexity and intensity of engagement with their yard spaces, were often reluctant to declare themselves gardeners, yet each pinpointed different material engagements important to gardening. In all cases, experienced gardeners considered essential the willingness to try and possibly fail. Within this sensibility is the knowledge that there will be the chance to try again, that, with the rhythms of season to season, and year to year, there are opportunities to experiment, fail, and succeed. "Well, I don't know that much, but I know just enough to experiment, just enough to get into trouble!" Tim chuckled, with his characteristically modest and understated manner. It turns out, as visits unfolded and I found out more about his childhood, Tim's parents ran a greenhouse business in a small town now at the edge of the Metro area. So he grew up surrounded by the cultivation of plants.

Barb, a former Hennepin County master gardener,[1] is helping her sister-in-law, Jeanette, reimagine her yard across town in Saint Paul. Barb and I talked about the process of helping a beginning gardener get comfortable and make decisions about their yard. "She comes here to my yard, and loves this garden, so that's what I think she's thinking. But of course, she's got really different challenges, and really different relationships with her neighbors, and skills or interests – so all of that, is – we're kind of feeling that out with each other. And I've been puttering with this yard for almost twenty-five years! And I've said to her, 'It just will take a long time.'" Barb asks my advice about how to help her sister-in-law see the possibilities of the backyard, and we discuss pros and cons for drawing directly on printed photos of the yard. "I think she just really can't *see* it yet. So we're working on that." Barb also wants to work with the established plants in the yard, such as an overgrown untrimmed old lilac in one corner. "She is very concerned about that, but it doesn't look all so happy now, but I think it's gonna give it a new lease on life." Barb is helping Jeanette work with what is there.

The attitude to try and possibly fail pervaded conversations about gardening practices. And it points to the ways the most avid gardeners often were the least attached to particular visions of what their yards and plants might be like. A specific creative vision seemed less important in most cases. Rather, participants who consider themselves gardeners understand iterations, attunements, and response at the fore of the ways they experience their yards.

This response relies on being attuned to the pace of plants. Barb tells me about a phrase she learned in the Master Gardener training program that has stuck with her, about the pace of plants: "sleep, creep, and leap." She explains, "You think about the first year that you put a plant in, and it's gonna just sleep. You're not gonna see much, it's gonna try to figure out if it wants to be there. Then the creep year – it comes back, it survives the winter, and it's gonna maybe send out a couple of new leaves, or it's gonna look like it might really grow. And then the leap year – after three years – you see it's really taken. And it lives there. And it's looking healthy and it's really gonna make it." Barb tells me about waiting to see whether an area she's planted with her sister-in-law with groundcover from a neighbour will really take off. "That's my dream, to have it be so lush and then you'll have to use the patio bricks as places to step between the plants. And if it doesn't work, it doesn't work!" Barb laughs.

In addition to the pace of plants, Barb tries to help Jeanette see the possibilities by looking at nearby yards. "You don't have to make a big investment. We can trade with people, we can work with what works in their yards." In this case, as soon as Barb and her sister-in-law were walking around the yard, a neighbour from across the street with an impressive garden came over to talk with them and offered all kinds of things she could divide from her own yard. In addition to this kind of awareness of what is thriving in nearby yards, Barb tries to help Jeanette see that things will evolve and change, and that moving plants will be just fine. "She asks me, 'Why would we move it after all the trouble of putting it in?' I try to explain that you kind of get a feel for it, and then things change – like the trees mature and you have more shade – but you can't know that ahead of time, what will be too shady, or where the plant feels like being." This is an example of how avid gardeners are attuned to the subtle and shifting needs of plants as they mature and develop across seasons and years, and also to the fluctuating resources available to them.

Geographies of Happy Plants

Whether or not plants are "happy" was a constant thread in gardeners' comments, used as a way of describing how plants might thrive (or not) in particular locations in yards. The sensitivity required to track these plant micro-geographies and anticipate how individual plants might respond to a change of location (and subsequent changes in water, drainage, nutrients, sun, or shade) constitutes a humdrum skill of little note for most participants I talked with. Helen, another experienced gardener in South Minneapolis, captured what other gardeners left largely implicit when she told me, "I've been surprised by how much time I spend just looking out, editing. Asking myself, *what do the plants need*? And I contemplate this and consider different things to do."

For Sandra in North Minneapolis, gardening in her yard is all about the ways her plants thrive at the intersection of touch and care over long time periods. She explained how gardening demands a certain degree of responsiveness to plants over time. And as she continued, the rhythms of this care come through as a rewarding, and deeply embodied, engagement. She tells me about her adjacent neighbours who invested quite a bit of time and effort into planting gardens in front and back about ten years ago but have never had the ability or the interest to keep up with them. "You know," she told me quietly as we stood along her front sidewalk glancing toward their yard, "I don't understand people who just *install*, and don't want to nurture these little things, and make sure that they're okay." Sandra brightens, "That's the part about gardening that is so rewarding, is that it's a living thing and it gives back to you. I don't care if it's vegetables or plants, but you're gonna love it, and it's gonna love you back by *producing*, and I think that's what it is about." For Sandra, plants are affected by the care people provide, and people are affected by the very growth and production of plant bodies themselves. All this happens at the pace of the plants themselves and their ongoing, and sometimes changing, needs for water, nutrients, and weeding.

Differences from year to year, as well as from season to season within one year, inform how people make decisions and make these kinds of responses to plants over time. Often participants recounted to me long and detailed narratives about particular plants – their origin stories, when they planted them, if they had moved them around the yard or divided them, whether or not they liked them, where such plants had been "happiest," or thrived the most clearly. For Kenneth, this takes shape in his relationship with one particular plant, a ladyslipper. Kenneth is an accomplished gardener and also a former nurseryman who is now a stay-at-home dad in South Minneapolis. He told me about the experiences he had had with the ladyslipper over the course of about the past decade, describing this as an ongoing responsive encounter between himself and the plant. The plant responds to its surroundings and care; in turn, Kenneth responds to the plant. He told me, "I do have a couple of favorite plants, so I always give them extra water. I protect them. One is my yellow ladyslipper. I planted it, and I didn't understand much about how it flowered. The first year, it flowered. The next year, it flowered. The third year, no flower. Well, then I read that it normally takes five to seven years before it will flower, if it's transplanted. And it went for four years without a flower. Now, it's been flowering the past few years, so I guess it's doing all right. But it comes out for a week, and then it's gone. And then it just looks like a weed. But I am pretty excited about this particular plant." Kenneth continued, "A lot of it is waiting. You experiment, and every year it's like, well, that didn't quite work, so I gotta wait until next year. I'll try something else and try something different. A garden, like a painting, is never really finished. A lot of it is just experience, people planting year after year."

Plants on the Move

One important dimension of this experimentation emerged again and again: people constantly move plants around. Participants often had their own philosophies about the best time to divide or move plants in response to changing seasons. In the fall, some gardeners like to make these changes to tidy things up and be ready for the following growing season. Others preferred to do this in the spring, feeling it made the most of the new growing season and kept plants under control. Either way, gardeners relied on their memory of what had happened in earlier years, success and failure, and could recount detailed micro-geographies for individual plants.

For those newer to gardening, trying to settle on how they want their yards to be, moving plants can become an obsession, even as it hinders the ability for smaller plants to become established. Jack recognizes this when he says, "I keep moving things around! Nothing can get bigger, because I keep moving it every year or two! But that's one of the things I like the best about my yard. Is making those changes. And splitting things up to share with other people." Jack shows me the "infirmary" in a corner of his yard, a small bed of plants that need some extra care and will eventually be planted in more high-profile parts of his front and backyards.

People also moved plants important to them from yard to yard, feeling anything from sentimental attachments to more pragmatic concerns about the money invested in buying them. Lorraine told me she moved plants in early spring, when she moved from a street undergoing major construction a few blocks from her current home in South Minneapolis. For instance, she tells me about a ground cover. "I had this in my boulevard at my other house, and it has beautiful little magenta flowers that bloom all summer. I gave some to my friend, and then she gave me some back when I moved. There's just a little spot of dirt out there, along where the trash and the neighbor's fence is. I thought if that grows back there, it'll look really pretty." In addition to plants and outdoor furniture, Lorraine brought a variety of objects with her, including a trellis, bricks, sculptures, and mobiles. She groaned at herself when she told me, "Ursula, I think I had more garden crap than regular stuff when I moved!"

Garden Memories

Beyond their material presence, past gardens and experiences with garden-ing are often right at the surface in affecting how people engage with their yard in the present. This comes through in the ways people understand what is ideal, normal, and valued; in bodily capacity with respect to the skills and experiences close at hand from which to draw; and in familiarity with the local biophysical climate and various plants. While taking a garden tour in a neighbourhood not far from my North Minneapolis study area (though distinctly different in terms of racial and economic makeup), I came across a poignant example of this kind of gardening in the image of past gardens. One resident has been shaping her backyard to mirror her grandmothers' gardens – one, a country farm garden, and the other a Latvian-American garden in the city. For this gardener, these remembered gardens give a rich sense of place and meaning beyond the present temporality. They also guide decisions about which flowers to include in garden beds; in this sense, the past gardens become static in their image, but they are also in flux through the ever-present processes of growth, change, and cultivation.

Usually past yards resonated in the study areas in much more subtle ways, often not surfacing until deep into a yard visit, a follow-up, or a story not immediately relevant, but related. Two dominant geographies of rural gardening experience came up in my study areas: childhoods growing up on rural Minnesota settler farms; and summer visits to extended family in southern states, retracing trajectories shaped by the Great Migration. Very often people with these experiences took for granted their gardening skills and capacities, or purposefully counterposed their current city yards with the utilitarian gardens of the farm.

Ann and her partner, Bonnie, have settled over the years on distinctions between front and back as territories in which each takes a primary inter-est. Ann considers the front her domain, and it is the place where she has slowly and methodically replaced lawn and existing hedges with a range of plants such as hostas and some annuals she enjoys and finds at nurseries. Ann navigates between space needed by the dog and carving out areas for her plantings. Bonnie maintains a gorgeous assemblage of blooms in the backyard. She grew up on a farm in rural Minnesota and credits this expe-rience for her abilities but also her desires for a garden that is now purely about flowers, colours, and textures – decidedly not food production. They spoke back and forth, finishing one another's thoughts, as we talked about their past experiences.

Bonnie chuckled often, sceptical of my project as she answered most questions, a reluctant participant. (Ann was much more interested to talk with me.) But when past gardening experience came up, Bonnie rose to the question despite herself: "I grew up on a farm. We gardened all the time. The grown-ups used tractors and plows to go between the rows of the vegetables. Because we canned for *weeks*. Dairy farm." Ann then chimed in with her dry humour, "And as a result, we don't plant vegetables." Bonnie added, by means of explanation, "Once I discovered a farmers' market – which of course we didn't have growing up – I was like, 'Oh! You can just buy these?!' So." She paused in her deadpan manner. "That's what we do now."

As the conversation continues, it's clear that Ann had very limited experience gardening before she met Bonnie. She has had to, in her words, "learn to enjoy the outdoors." They both relish telling me a story about a trip to an aunt's house, where a big garden included potatoes. "The kids now haven't a clue. We took her nephews to the lake once and told 'em to go pick some potatoes or something. They came back and said they couldn't find anything! They didn't know to look underground!" Ann interjected, "That was ME!" Bonnie continued, "Well, see I just grew up with all those things. I mean, I can spot wild asparagus in the ditches." By contrast, Ann's experiences growing up "in a quonset hut in university housing" included few opportunities or obligations to cultivate plants. "Well, we had a little green patch. And there was somebody kitty-corner from us who paid us a penny a dandelion, and his daughter and I would dig them up. But, mow? Plant? What? So, yeah. I never, never, *never* picked a vegetable of any kind. It's an acquired taste."

Cultivation is also a skill informed by these early years. Over time, Ann has adapted to living with an experienced gardener and has taken gardening on herself by slowly taking charge of the front yard territory. "Gradually, I started sneaking hosta in. We fenced the front yard because of our dog, and grass hardly grew there under the big tree. I pruned back the honeysuckle bushes, and then I spent a year sifting the dirt, getting the roots out. And then I thought if I put a few hosta in, she wouldn't notice. I'd dig up another row of grass and put some plants in. And Bonnie's very observant, but not necessarily apt to comment." There seemed an implicit agreement between the two that this front area was Ann's place to mess around, and that Bonnie's gardens in the back were a more serious and skilled endeavour. In describing them, Bonnie tells me her motivation is recreation, but she also says she might have been a farmer if her grandparents hadn't sold the farm. She continues, "It's something real to do. It is unlike office work." And Ann adds, "And you like having the cut flowers. You're more wanting the English cutting garden. I like that, but I don't have any of that planted in my brain. I'm the shade gardener, so I buy hosta based on a good name, and a good price. And then, I just put 'em in the ground and see what happens."

In ways similar to Bonnie's childhood experience with large farm gardens in rural Minnesota, other participants had experiences learning about large farm gardens through family roots in southern states. During the Great Migration of the early twentieth century, African Americans moved from southern states to northern industrial cities. This geographic trajectory was a fairly common thread across African-American participants in North Minneapolis. Often, they would visit family in agricultural settings in the South, where they were exposed to the pleasures and labours of large gardens and farming. The visiting children learned how to work in gardens and how to pick vegetables and fruit, and they helped with tasks like canning on a large scale.

Through stories of childhood yards in Chicago and summers in Alabama, Sandra in North Minneapolis talked with me about the importance of creativity in learning these kinds of skills. She plants roses in honour of her mother, who loved raising roses in their Chicago yard. Sandra framed learning these kinds of cultivation skills through family as one means for survival and thriving as African Americans, in response to ongoing obstacles.

Elusive Garden Dreams

What to do when all these yard and gardening activities seem impossibly hard? When someone didn't grew up helping with any gardening or take much interest in plants. When they are not set up with the right tools and don't know what to plant, or where. Plants wither, leaves curling up, and eventually die. Although the people who chose to be part of this study were likely more adept at gardening than a cross-section of all Minneapolis residents, I tried hard to find participants whose yards were not necessarily spectacularly cultivated or whose own skill and sense of gardening identity fell more into the novice or uninterested end of the spectrum. Among these participants, several felt that they were failed gardeners, despite interest and effort. One of these is Sheri, who, from our first conversations, told me she felt she has slowly killed the nicest plants they inherited from the previous owners. She and her husband's yard is largely lawn, with some foundation plantings, and a small vegetable garden in the back. Her gentle and some-what tentative nature became even more hesitant once we were outside, walking around the yard. Sheri describes herself as a hopeless gardener, and she pointed out many areas around the yard where she had tried, and failed, to plant various things, or where she was unsure about what to plant. Yet she still talked about finding respite in the vegetable garden, especially the way going outside to pick fresh mint leaves for tea punctuates her day. She tends her small vegetable garden and a few annuals in the front yard, and their teenage sons mow the grass, and that's about it. Sheri points out mari-golds she has planted below the front picture window but calls them by the wrong name. "See," she says, "this is where I've tried. But I don't know ..." her voice trails off as she leans down, touching the little plants and drawing my attention to them. Despite all these struggles with her yard, somehow Sheri seems determined to keep trying.

These kinds of "awkward encounters" open up thinking about how peo-ple feel a deficiency or lack in their relationships with their surroundings,[2] and about how someone like Sheri understands and feels this lack. Her experience is full of uncertainty, a kind of fatalism that she jokes about mixed with pride at attempting to grow annuals. She relies on a neighbour's wisdom and knowledge about plants, as well as her willingness to share plants themselves. Again and again, Sheri credited her neighbour Fanny with guiding her toward any success that Sheri has had.

This kind of learning from neighbour to neighbour happened throughout study sites, and often participants credited adjacent or nearby neighbours for getting them started and seeing them through difficulties, projects with particular plants, or yards in general. Two of the most accomplished gar-deners in the study areas were not always so. Barb and Marta have been neighbours for more than forty years. They know one another's yards, and they know one another's ways of being in their yards. Marta's yard is filled to the brim with blooms – especially her south-facing side yard, which is adjacent to Barb's study windows. Barb told me that this close proximity and the experience of Marta's engagement with her yard have been central to her understanding of what it means to cultivate her own kind of growing with her yard.

> I think her garden is an extension of her love and her personality, so the garden becomes – what you walk up to is what she wants you to *feel*. It's really beautiful, it's really neat. I have a different aesthetic for gardens, but it was good to have that influence, where you – where you tended it, and you enjoyed it, and she was out in her garden. We would talk and we would visit, so I got to see up close more like the management of the garden, the enjoyment of it. That you don't just plop in the plants. It's like children – you live with it and you work on it.

Cultivation is thus a purposeful and creative practice. This account of Barb's points toward learning from Marta the ongoing formation of a garden through constant engagement. Such learning is one result of an embodied engagement with yards and illustrates how cultivation takes on multiple social meanings and a variety of material forms over time. Cultivation is full of imagining and experimenting. I insist that yards are more than gardening, but it is clear that, by their very nature, cultivation can be central to many experiences with yards. Cultivation is embodied, learned, and forgotten, changing in relation to bodily capacities. Cultivation is also iterative. Aspirations to make new worlds call forth skills of care and entanglement with others. How will new worlds emerge from the ground?

Rhythms and Time

Keeping Up with the Pace of Weeds

Weeds provide a focus around which some of the above cultivating practices and experiences orbit. The *pace of weeds* means the particular temporality of those plant bodies that gardeners (and others) attempt to eradicate from their lawns, gardens, and yards. Throughout this study, weeds and the weeding activities that usually accompany them – or the experience of obligations to weed – were a constant source of reflection and conversation before and during yard visits. The very visibility of weeds, especially in front yards, instigates this kind of maintenance – pulling, digging out roots with tools – and a presence always there, only sometimes more or less at bay. Whether people love or hate to weed, a language of "keeping up with" reveals different ways of experiencing time and being in yards. The pace of weeds had a double meaning: on the one hand, the persistent growth of the plants themselves; on the other, an attunement to human surveillance through people's ever-present awareness of others' perceptions of weeds in their yards.

Perspectives on weeding often fall into one of two distinct camps. For many, weeding is a loathed, but necessary, obligation or duty. These people despise weeding more than anything else, especially the sense of crushing obligation and feelings of failure when they weren't able "to keep up with the weeds." Other participants love weeding above all yard tasks and re-ported feelings of peace, calm, and meditation in the practice of weeding and in reflecting on weeding labours after the fact. Kay in North Minneap-olis spoke about weeding as a kind of time outside of time: "I lose myself." Likewise, Barb said now that she is retired, hours can go by without her even looking up from her weeding. Her husband, John, said he's taken to carrying her purse inside for her from the car, because even the short walk to the back door can take several hours if she bends down to pull a weed. People had favourite tools and tricks. John in Northeast Minneapolis uses a Chicago Cutlery knife with an aging wooden handle. He demonstrated how he pries up the long white thick roots of dandelions. When he was done, he nonchalantly tossed the knife at the lawn, where it stood, handle up, for the rest of our yard visit.

Changing Bodily Capacities

Often keeping up with yard tasks like weeding involves changing bodily and social capacities, such as asking for help from neighbours. Lorraine is a woman in her early seventies, retired, a mother and a grandmother, now living alone after losing her husband to illness several years ago. An expe-rienced and skilled gardener, she has been adjusting to increasingly limited physical abilities to bend, stoop, dig, carry, and perform other kinds of labours in the cultivation of her yard. I asked her how she would describe her yard and what she does there. Without pause, Lorraine focused on her backyard and told me, "The back feels to me like a certain place of retreat." Then she paused. "Working with plants for me has always been therapeutic," she said. "When I was working [as a minister], I would come home and be exhausted, but I'd go out and garden and it'd be just like taking a nap. Sometimes it's challenging because I now have two hip replacements, and I can't squat down – there are certain things I can't do. So I've had to adapt, and I'm going to be making more adaptations." Lorraine paused, then continued in a strained voice, "Which is really very frustrating for me." She brightened a bit with some effort. "But mostly, my backyard is my refuge and my sanctuary."

Lorraine did not dwell only in the positives about her yard experiences. She explained during the course of our first yard visit that recent challenges with arthritis significantly limit her own physical capacity to do the things she wants to around her yard. She was frank and open about the challenge of accepting these changing physical limitations, and I could hear the frustration in her voice at several points – for example, as she flatly described her desire to accomplish more in a given day and the reality of needing to stop because of sensitivity to heat or pain. Because of these physical limitations, she has relied more and more on the physical capacities of others. This took shape in the ways she depends on her son and daughter-in-law, who live several houses down the block, as well as her neighbour immediately to the north, Kenneth. These names came up again and again as Lorraine told me about changes she made to the yard over the past three summers she has lived in this house.

At the same time, the practices Lorraine called "tending the yard" – weeding, mulching, deadheading, watering, seasonal tasks like tidying up raspberry bushes or dividing and transplanting perennial plants – the many different labours of yard work, also give her a rich source of pleasure. "What do I enjoy the most about my yard? Hmm, probably sitting and enjoying the beauty that I've created, and then next would be just *actually doing* the yard work and making it look really pretty." Her yard is a place to be tended and enjoyed, a place where her own embodied relations with her material surroundings impinge and shape how she finds meaning there. And the combination of her visions of gardening as she wants it to be with the limits of her changing bodily capacities invites and demands particular social relationships with her immediately adjacent neighbours, and those along her block.

This need to keep up with gardens, plants, and yard maintenance, in conjunction with changing bodily capacities, is also captured well by Jack. In his experience, the size of the yard and expectations for what is gardened make all the difference. "I think a lot of people don't have a lot of time, to maintain, and a lot of times, I think they maybe have too much garden and it gets overwhelming, and you gotta be really careful about that. Cause your garden can kind of overwhelm you, and kind of own you and run you. And, like, who wants that?" The past few years he has been helping his mother, in her late sixties, more and more with her large suburban yard. "I look at my mom's yard and at her age, it's kind of overwhelming. And she's decreased it considerably over the last twenty years, which is a great idea. And I think people just take on too much, or get kind of nervous, and then it's like making your first vegetable garden – you plant too much and then it's like, 'Oh gosh, now I don't want to weed it, water it, and this and that.' You have to watch that or it'll run your life and you don't want that." In contrast, Jack has been happy with the small size of his yard and tells me he can keep up with all the necessary tasks in one long day. Or he simply spends a little time each day, in the morning. His situation is a bit different from most, in that his work is quite independent and the summer season is slow. "Usually I wake up and I go outside and I water. And that's kind of what I do. I do that, and I do a little weeding, and that's pretty much all – you know, that's about forty minutes a day, is all I put in. Sometimes I put in more."

Participants – and not always only the most experienced or elaborate gardeners – often talked about such yard tasks, and in particular gardening, in terms of "the doing" and "the having done," both of which involve activities, skills, socialities, challenges, and engagements with more than human surroundings. In many cases, the labours and pleasures of gardening establish a setting for simple activities such as sitting with a hot or cool drink, looking and watching, listening, being in and of the landscape for a period of time, or simply being out of doors. Sometimes these are ordinary and unremarkable times, constituting rhythms and patterns in daily life. Or they make up unusual moments, exceptions, or temporalities outside the confines of linear time. The doing of yard tasks and then appreciating (or being frustrated by) the results once they are done are important in the ways people understand their own cultivation practices and those of others.

All of these particular weedy rhythms, and the diversity of experiences with them, point toward the importance of embodied engagements with yards and yard socialities. These embodied engagements indicate the importance of maintenance within inhabitation – shoring up, keeping up with, making do, tidying up, keeping chaos at bay. For yards, people keep up with neighbours, appearances, plants like weeds, trees that drop leaves in fall, plants that benefit from dividing and mulching in spring. These are tasks done for others, and also for the self. Rhythms of work and enjoyment. The satisfaction of accomplishment. Inhabitation embodies these unseen small iterations in surroundings each day, month, year. This activity of inhabitation is the ongoing formation of urban environments and the glue that keeps things together, shores things up, patches and repairs.[3]

Experiencing Tree Time

Trees become revealed as a measure of time through their sudden destruction and absence, as in the case of the tornado discussed in the previous chapter. Trees also endure and change over long periods of time. These are less dramatic changes, but no less meaningful. In Northeast Minneapolis, Tim is telling me about the trees in his yard during our first yard visit. There are about five maturing oaks that he planted around twenty years earlier and are now a respectable size. "I guess you could say I almost feel like these trees are my children, in a way. I just like watching them grow and get bigger." Tim has a quiet and calm demeanour. He is patient with me and interested in participating in the project but not overly talkative. He is in his sixties, the first summer I meet him, married with no children, and he has lived in his house for two decades. Tim has limited mobility in his legs and uses a motorized wheelchair. His yard is neatly maintained, with a spacious side yard comprising mainly lush lawn dotted with trees, with some small planted beds around the margins like the front sidewalk. He and I have a particular connection, and I feel my questions stir something up inside of him. I sense right away that my quiet seems to match his quiet. When I see him for the second yard visit the following late spring, he tells me, "I remember our conversation like a therapy session." He remembered telling me all kinds of things he hadn't said to anyone before then. On the spectrum of yard visits, his was not remarkably personal or open, but it was clear the encounter felt that way to him.

The question of trees has struck a particular chord with Tim, and in his voice I catch more emotion. Over the course of two summers, as I visit Tim's yard periodically, he discusses the trees and how they are doing, any changes. He primarily trims the trees himself and has hired a tree-trimming business only once or twice, when the limbs have grown up beyond his reach. From the deck at the side of the house, Tim likes to just look out at the trees, watching them grow.

Tree Proportions, Scale, and Measures of Risk

Trees also serve as measures of scale and proportion in yards. For John, an artist, former cab driver, raconteur, neighbourhood organization member, and resident in Northeast Minneapolis for over twenty years, the importance of yards is constituted primarily by physical elements in proportion to one another. He seems to have in his mind's eye a sense of the appropriate physical elements, how their relationships should constitute yards, and how they are arranged and delineated in the neighbourhood. In addition to fences, which were a major preoccupation of his, John talked at length about trees in his own yard and in general. John's experiences with trees are inflected with a deep reverence but also focus on the management of risk and potential destruction to houses from falling limbs.

When John moved into his house twenty-two years ago, one of the first projects in the yard he undertook was to make a rectangular brick surface in his side yard about ten-by-six feet, and to begin building a scale replica of Chartres Cathedral, about six-feet high. The project took about nineteen years, he told me. He made the cathedral out of scraps of plywood, two-by-fours, and other lumber odds and ends, and he ended up painting it blue to protect the wood from weather. As he tinkered away at its construction, people would stop and peer at the model through the hedge and fence along the front sidewalk of the yard. Living between two popular neighbourhood bars, John told me it was always interesting to listen to what was said when groups of inebriated people would pass by on weekend evenings. The nonchalance with which he told me the story of the cathedral's demise belied the attachment he still clearly feels toward the structure, expressed in his affection for the remains of it and also the detail with which he recounted building it and accidentally destroying it. He took me into the garage to show me a practice spire he had built as he was still getting the proportions just right. "This one's *nearly* perfect," he told me, as he touched the very top of it. John built all of this by eye, experimenting with size and shape, angle and construction. He says, "It's easy, all of this is from free wood, free lumber. You just try it out and eventually it works." John tries to insist to me that nothing about the cathedral project is unusual or required special skill. In the nonchalant way he discusses this and his other projects, John insists these impulses to observe and to make using what is at hand are latent in everyone. He also recognizes with some wistful sense of loss the unequal experiences of constraints on time, pressure to work for survival among working classes, and the loss of skills that would enable people to build what they might dream.

The Chartres model he had built for nineteen years was accidentally smashed three years later, in a matter of seconds as he cut down a large box elder tree. Pieces of the cathedral hang along the back fence behind the garage, not a visible feature of the yard from most vantage points, but still present in the background. Traces of the tree remain, too: one large chunk of branch or upper trunk lies in the side yard, while the wide trunk, still about twelve feet high, rises up out of the backyard. New shoots grow from it, which John trims back periodically. "It's neat to watch it change over time," he told me as we looked at areas of the trunk that were disintegrating on the ground. "The thing about trees is that they are going to fall down. The branches are going to fall down, and you don't know when. But your house is going to be in the way, and then you will have problems. Like my neighbor, I keep telling my neighbor, those branches are right up close to your house!" John's insistence about the dangers of falling tree branches has meant that, over time, he has methodically trimmed back most of the trees in his yard. He scoffed when I asked if he hired anyone for these jobs. While John considers trees important elements to yards, his own relationship to the particular trees in his own yard takes shape through the structure of his house, and the possibilities – in his mind, inevitability – of damage. For John, the trees embody unpredictable forces, which can be constrained only by proper and vigilant management.

Time and rhythm determine how we live with surroundings day in and day out. This living-with unfolds over multiple interfering temporalities, over short and long periods, over lifetimes. We can illuminate the significance of this sense of ongoing formation and openness by applying Lefebvre's critiques of modern capitalist social relations. As discussed previously, Lefebvre argues that everyday lived experience is the key to socio-political transformations in conjunction with built environments. These experiences resonate with the ongoing practices of formation in the making and growing of yards. When we apply Lefebvre's project of *rhythmanalysis* – attending to the multiple temporal registers at play in the production of any given spacetime – to yards, we see how the pace of plants, routine habits, and rhythms of people's lifetimes course through their diverse engagements with yards.

Being with Others, Being Out of Doors

Encounters of Chance and Intention

We walk down the sidewalk in North Minneapolis – myself, in my thirties; our guide, Kay; and my colleague, in her sixties. I am getting the lay of the land, before any in-depth yard visits, hanging around and getting to know people. Kay shows us around by going from yard to yard along her block, a shared green space, and several distinctive yards several blocks over. It's springtime, with fresh light-green leaves and promises about how things will unfold. Kay has lived on the block for more than twenty years, and she walks us through her neighbours' side and backyards, showing us hidden spaces, utterly at ease. She introduces us to several women along the block – one inside cooking – on our way to the shared community garden. Kay led us along their sidewalk to see the back. She said, "You can see this is a mature garden, established, old." Kay brought us into several more hidden oases behind and beside homes, with noticeably inward styles of gardening.

Although rich and full with maturity and care, the gardens also felt aesthetically quite restrained. In yards of South Minneapolis, growing trends of native species and food production meant blocks were often interspersed with a certain kind of unkempt performative shagginess. But here, the

relative biophysical restraint of borders and rows prevailed. This was counterposed with the ease and familiarity both between African-American and
white neighbours and – extending out across all kinds of apparent social
differences – toward passers-by and neighbours further afield. I felt this the
first evening there and on many visits afterwards. This familiar ease was
evident in the calling out of names and greetings down the block, talking
in back alleys, long conversations in chairs outside – about politics, history, and equally about nothing much at all besides watching kids grow up
and enjoying others' company for years and years. On that first walk, Kay
shows us a house across the street that was finally condemned after some
drug use and other violations, and nobody was sure it would be habitable
again. It was a large, solid home like most of the houses in this area of North
Minneapolis, but now with run-down plaster and brick, windows not entirely
intact, and an orange notice on the door. Its presence with dark windows in
the early spring evening brought some complexity to the otherwise warm
and social circulations of these established yards and neighbours. It also
was a shadow of the significant disinvestment this part of Minneapolis has
experienced since the 1960s.

Several miles to the south, neighbours along one block gather in a backyard
each Friday evening from May to September. The location is announced with
a repurposed political yard sign in the front yard declaring "TGIF" (Thank
God it's Friday) and indicating the gathering will be in the back. The rules to
maximize enjoyment and minimize pressures to host or outdo one another
with a formal affair include: no food or snacks unless they come directly from
a bag (like potato chips), no alcohol costing over ten dollars (no fancy wines),
and nobody allowed indoors, so no cleaning is required (everyone can walk
back to their own houses to use the bathroom). Kids are underfoot and play
in the driveway, older kids move around front yards along the block as their
parents socialize in back. People filter in and out, gathering on decks and
back patios, standing and sitting. Attendance is loose and fluctuates with
the summer season. Notices are sent out via a block email group. People on
neighbouring blocks know this street for this ritual, even inciting a low-key
competition, with a similar rotating gathering on a different day of the week
for the block directly to the south. At the first gathering in May, old timers
introduced and welcomed new families to the block, as people caught up with
longer conversations than passing hellos after a long and snowy winter.

This dimension of inhabitation is about sociality and connection. Different neighbourhoods have different socialities, or ways of being social. Spaces such as yards, especially as they are in between indoors and outdoors, figure in these socialities and the ways people encounter one another. These socialities are informed by a particular position within broader urban geographies, including histories of investment, public services such as transportation, destruction during urban renewal, and contemporary pressures for redevelopment leading to gentrification. Often enacting deeper social connections through yards requires only a small nudge, and people start to show up, to start something more significant. In examples from North and South Minneapolis, yards become activated through encounters. While styles might be different across the city, these North and South Minneapolis neighbourhoods share the impulse to experiment and use their yards as social spaces.

Out of Time, Out of Doors

Inhabitation entails not just active cultivation, with all of its embodied practices, like planting, designing garden beds, digging, and weeding. Inhabitation also involves reflection, which takes shape through practices of sitting, looking, and thinking. In this section, I examine some of the primary ways in which people discussed how they inhabit yards beyond gardening. These stories highlight the necessity for pause points and rest, which together allow for sensory attunements to "just being outside," as many people told me. Here, yard rhythms are about slowing down, pausing, resting, and noticing surroundings. A rhythm of getting out of time.

Cultivating plants certainly contributes to these experiences through the shaping and reshaping of these spaces. Avid gardeners, in particular, talked about sitting in their yards and admiring their labours, as well as appreciating the growth of the plants themselves and other organisms in yards. But gardening is certainly not necessary for such sensory engagements. As can be seen in the vignette about Michael at the opening of this chapter, even the barest and most "empty" of yards can hold value and meaning as a place to experience. For Michael, sitting on the front steps is a means to inhabit social worlds of his neighbourhood – people passing by with dogs, neighbours coming and going. In a broader way, yards become places, or settings, for people to pause in order to reflect and be. In this way, they are settings that afford people the capacity to dwell.

This kind of inhabitation most often included simply *sitting in place*. Many people told me they spend time contemplating their yards from favourite vantage points – often they talked about having morning coffee in some proximity to their yard and using this time to figure out their day. For Barb, an accomplished gardener, the cultivation of her yard provides immediate pleasures in just sitting and being in the space. Furthermore, she relates this to care over successive lifetimes by residents of a single house. In this way, the cultivation of her own yard provides a way to provision future inhabitants with what she considers a meaningful setting – she is cultivating this yard as a future timespace. And this attitude resonates with the ways she understands provisioning newcomers to the neighbourhood with plants and knowledge. She tells me she thinks about all these things as she sits in her favourite spot in the yard.

Where people sit has a lot to do with the basic elements of light and shade. Not surprisingly, when people spend time in their yards, they tend to seek shade in summer and full sun in spring and fall. The properties in all the study areas of this project are laid out on street grids that reflect the cardinal directions. This orientation determines when and where in a yard sun or shade are available at different times of day and times of the year. Distinctions between front, side, and back depend on the rising and setting sun but also the built structures and features of each yard. During the field seasons of this project, the desire for shade was particularly acute, as the summers were full of hot and humid days. Shade determines much about yard furnishing and use, especially positioning furniture and the use of patios, stoops, and porches.

In all the study areas, then, being on the east or west side of the street allows different socialities to emerge at different times of the day and year. For instance, Tina's front patio project, for which obtaining permits was challenging, provides what she calls "usable" space for summer breakfasts and time to be outside in the summer. "In the summer, we're out front in the mornings for breakfast and coffee. Then it's opposite, in fall and spring, when we sit out back on the deck for dinners, where it's shady." A few blocks down, and on the opposite side of the street, Leslie told me about realizing they *could* put chairs in their front yard. "I don't know why," she said, "but when we lost the shade in back because of the old tree that had to come down, we put two chairs in the front yard for the first time, and it was so nice to be out there in the evenings! We found we talk to neighbors more, or people walking past. Just little hellos, but it's been really nice." Throughout the study area, participants made this same point.

Yard inhabitation has to do with being out of doors and being with others – human and non-human. Living with yards is becoming attuned to others through relations of response. This ranges from differences in ways of being social through yards to individual experiences of sitting, listening, looking, smelling. Living with yards entails the pleasures of noticing, pausing, and simply being outside.

Conclusions: Living with Yards

There are many yard stories in every city block. Throughout them all is a sense of ongoing formation between person, experience, and space. I call this *living with* to highlight the emphasis on time and relations. Living with unfolds across architectural forms, urban designs, and histories of development. It is how people live. Living with yards captures these relations from purposeful creative reimagining of space to quietly sitting on a front step. These are the day-in and day-out activities through which cities are made and remade.

Throughout all of these stories and experiences, we can ask about the political and social aspects. Who has access to space and time for these experiences (and when and where)? Participants' income and home mortgages range, and some people clearly have difficulty finding resources for routine maintenance and upkeep, while others embark on or have completed large-scale renovations or regularly invest in plants and gardening supplies. Likewise, who has the experience and family histories to inform skills like gardening? It is clear from these stories that living with yards entails particular privileges and responsibilities, informed in this context by racialized urban geographies. Does this change the way we understand urban geographies of nature? This partial and situated slice into yard experiences offers an opportunity to consider the possibilities of a recuperated phenomenology of environmental care.

3

CHASING
YARD
AFFECTS

An Invitation

This chapter is a photo essay that offers multiple and partial views into yards and yard experiences. I pair image and text as an exploration of how the one exceeds the other. This essay is one answer to the challenges of how we might find modes of address that are adequate to ordinary circulations of affect, power, difference, and care. It is also a provocation to consider how images, research praxis, and writing can be interwoven.[1]

This essay is one attempt to "remap the familiar,"[2] and also to learn from yards and what people are already doing as they live with these spaces over decades. Photographs provide perspectives, information, feelings, and encounters distinct from other research registers. The unfolding diversity of spaces and perspectives invites serious reconsideration of environmental care (or challenges to it) beyond that formally sanctioned by contemporary green trends or best practices. I intend for this experiment to make more room to reflect on conventions that prioritize written representations, in an effort to get more from visual ethnography and the making of images as part of research and environmental planning. This visual register will not be new for designers, but perhaps a slower pace of looking and feeling how people live with their surroundings will be a bit new. As I have shared this part of the project over the years, colleagues in the social sciences have been unusually interested in such an approach. Even as we are increasingly immersed in complex and primarily visual digital worlds, with this chapter I argue that the possibilities within relatively simple relations between image and text on a page have not been worn out. What might image and text do?

This essay is made from the interaction of four interwoven voices, each delineated by different formatting on the page: images made by me, participants' quotations from interviews and yard visits with me, my observations about the role of images within my research practice as well as excerpts from research notes, and guiding fragments from others about key concepts.

The images in this essay have paused, sped up, framed, constrained, instigated, generated, contradicted, and validated representations of yard experiences. In none of these functions are photographs neutral or straightforward, nor are they intended here to illustrate or record an objective truth. Such views ask us to more deeply consider the possibilities of what our surroundings might do – in cities, in neighbourhoods, in daily life, with others, and for others. Most of all, through this essay, I invite you to be affected by these yards.

Where shall we start?

I JUST TAKE changes as they go, start with one little area, and do some changes there. This summer, I moved some bushes around – a spiraea, given to me by my girlfriend, moved from her yard. So I planted one on each corner of the house. To do that, I had to move other things around. So I also put in this shrub.

Many times when I asked people, "Show me your yard," a kind of inventory unfolded as we walked. This was important – in letting me know what they thought I wanted to know, but also what they wanted to tell. Some of these accounts were clearly familiar and even well rehearsed. Others emerged in the telling. The plants especially were a way to focus our eyes, our conversation, our gestures.

Touch was significant in the telling of these plant origin stories. Where, when, under which circumstances human lives and plant lives came to be entwined. Distinctions were made between plants, differences noted, touched, pointed out. People and temporalities were also woven into the descriptions.

Required of me: most of all admiration, nodding along, agreeing (sometimes too much), questioning, documenting.

The importance of touching yards comes through every time people leaned toward plants, brushed their hands over them as they talked, ran leaves between their fingers, pointed and pulled foliage aside. They cradled blooms in their hands and pulled weeds from their footings.

Absent-minded or purposeful, while they talked, some hung on to fistfuls of leaves. Plants brushed our shins, we ducked under branches, we looked up with mouths open, admiring the shade from a last big elm canopy. Hands searched to locate and point out a particular plant among others.

Touch is part of a multisensory and embodied inhabitation of yards, including smell, taste, sound, sight. Touch was particularly important throughout yard visits, as residents showed me plants, and it is implicit in many tasks and activities around yards.

WELL, IT'S ALL ONE. It's all kind of *my* . . . What to call it? My whole. Yeah, my comfort zone, this is mine, this is where I belong, this is what I take care of.

Studying Home Grounds

Studying home grounds is an intimacy of engagement. As show and tell, as host and guest. The experienced with a place, and the inexperienced with that place. I asked people simply, "Show me your yard." Then I followed their lead. What do participants think I want to see? What do participants think I want to know? What do they point out, and why? What do participants say, and what do they do? Being outdoors as bodies, together. Senses, labours, attunements – this was a ballet at times, how we moved through yards, deployed touch, always leaning, pulling back, cradling blossoms in hands, smelling, admiring.

Yards invite, demand, and overwhelm certain relations of attunement and practice between the capacities of bodies and surroundings. If these affective dimensions of yards are felt as so important to people's experiences, how do we find modes of address adequate to their natures? Circulating in and through bodies, rhythms, and places, affects have been considered an excess, spilling over the concepts and categories through which representational thought and language know the world.

Ethnography can be particularly well suited to studying affective dimensions of people's relationships with their surroundings, because it is all about being immersed in others' time and space. The practice of studying affect requires an attunement to relations of response. The most powerful ethnography demands that a researcher cultivate an openness to being affected, in itself a philosophical orientation, and a willingness to try to see and reflect on the affective power of one's own presence on the people and surroundings in which one is immersed. Furthermore, sharing time and space with participants and places means that embodied engagements with surroundings necessarily become part of the world of study.

ISN'T IT BEAUTIFUL HERE! I need this emotionally somehow. It does something to me. It cheers me up. It makes me feel happy.

Marta's side yard is a neatly maintained and joyful place, full of colourful blooms. She grew up in hard times during the Second World War in Germany. She was the fourth of eight children. Her grandparents' farm was a respite during these years. Marta told me that she took a lot of pleasure in the physical work and said she would sing and sing in the fields while working – that others would notice this about her.

MARTA / YARD DIARY / 10 JULY
How do you feel about your yard today? OR Write down any additional comments or thoughts below.
It's blooming to my satisfaction.

MARTA / YARD DIARY / 18 JULY
How do you feel about your yard today? OR Write down any additional comments or thoughts below.
Just glad the ground got a good rain.

Marta loves to work in the midst of the flowers on the south side of her house. She told me it gives her a similar feeling of freedom and openness she remembers from working on her grandparents' farm. Sometimes she brings out a chair to this part of her yard. Sometimes just walking through it is enough.

IT'S FUN to have this [a raised garden with vegetables]! I like it. This is the first year. I just haven't done as much for it as I should have. But I've eaten from it. Collards and broccoli. And here are carrots. And I had an eggplant. It was a tiny little thing. But I ate it.

These greens look good. I think it needs more water, though. I haven't been doing it diligently, but I do come out – I should water it today.

The relatively new lumber is light coloured and fresh from this season. Tags are lined up with each row, the words mostly worn off from sun and weather. The raised bed came from a local organization focused on food security and environmental justice in North Minneapolis.

I'M SO EXCITED because this is my first year with tomatillos and they're in there. I just can't wait till they explode. One of my friends, she said that tomatillos are like the Incredible Hulk of the plant world – cause they just bust out of their husk.

THIS IS WHAT I DO when I'm talking to anybody, on the phone, you, whatever – I weed, I tidy up, I can't help it.

It's the garden that brings me out to the front. Just to keep it nice looking because it's the front, and I want people to enjoy its weirdness. This was one of the first things we did that first year we moved in. When you first cut a garden [into the sod], you have to keep cutting it, pulling the grass that tries to grow, I have kept up with that.

I imagine this will keep someone from buying my house someday [because of the gardens and necessary caretaking]. You know? Like they'll go, no way! But I'm not going anywhere right now, though I don't plan on staying here forever; this is not my home, this is where I live right now. I'll go back to the east coast. But this is a good place to be for arts, so it's a hard decision.

HELEN: I love it most in the spring, when everything looks so fresh and green. And it's that really beautiful light green. Now everything just looks a little tired. But it's a good time of year for planning. That's – I think – one of the things about gardening that has surprised me. I was sort of surprised about how much thinking I do. About it. I'll sit out here and I'll think, "It *needs* something. What does it need?"

The whole process is so – um – amazing. Like in February, you look out at the landscape and you think there is no way those tulips are gonna come back up this year. And then they do. So it's this whole birth/death cycle, it's really kind of surprising, and wonderful, every year.

URSULA: Are there particular aspects that have surprised you?

HELEN: I was surprised at how inaccurate the word "perennial" is, and how much – when people say, you can just plant a perennial and forget it – how wrong that is! Because right now, I have divided a lot [of plants] this year, but I have a list of all the things that need dividing next year. That really changes the look of a garden, because it goes from being overgrown to looking like you just planted it all over again. The whole process is never ending.

It's been fun to do it this summer as a partially retired person. Because I always thought, if I could just spend an hour or so a day, I could really keep this under control, especially weeds. And that does make a huge difference. And it's very enjoyable to go out there in the morning, and work.

And Joe and I see it very differently. We'll be out here [on the back porch], and Joe will say, "The garden looks beautiful." And I'll say, "You know, it's got a lot of weeds, I really don't like the way the iris are looking." So we enjoy it in our own ways. He thinks mine is too negative.

JOE: I like the process. I like the cyclical process. I do appreciate the changes Helen makes after they are made. But digging all the holes and planting stuff isn't – well, sometimes it is fun. Sometimes I enjoy it. But a lot of times, it's just nasty. Most of what I'm doing is working out there. But that's still enjoyable.

THIS IS A KOREAN LILAC that was here [when I moved in], and oh my gosh, it smells so good! Right by the kitchen window – it's just wonderful! And it's late blooming.

I just love this funky old gate. I just like it cause it looks old and you know, I didn't have it replaced when my son bought the new fence for me. And just beyond it is a trellis my kids gave me for a gift. I moved it from my old house. I just bought a clematis for it.

I love just this whole gestalt of this part of the yard. I mean, I just kind of like little rooms in the yard, and this is like the entry and the outlet part on this side of the house. The way into my backyard oasis.

Bodies in the landscape, of the landscape.

Sawing is one of a suite of commonplace tool-assisted activities, including also hammering, pounding, and scraping, that all involve the repetition of manual gesture. Indeed this kind of back-and-forth or "reciprocating" movement comes naturally to the living body. In a fluent performance, it has a rhythmic quality.

This quality does not, however, lie in the repetitiveness of the movement itself. For there to be rhythm, movement must be felt. And feeling lies in the coupling of movement and perception that is the key to skilled practice.

By way of perception, the practitioner's rhythmic gestures are attuned to the multiple rhythms of the environment.

Rhythm, then, is not a movement but a dynamic coupling of movements. Every such coupling is a specific resonance, and the synergy of practitioner, tool and raw material establishes an entire field of such resonances. But this field is not monotonous. For every cycle is set not within fixed parameters but within a framework that is itself suspended in movement, in an environment where nothing is quite the same from moment to moment.[3]

The capacities of bodies are constantly changing. As bodies grow with landscapes, each makes new capacities. Tasks that once were easy become hard. And challenging skills become habits, like second nature, over time.

All of this cultivation is like an offering people make to one another, to plants, to animals, and to themselves. An offering of wild blooms, colours, and textures, which are alive.

Seeing Familiar Terrains

An image gathers together forces, making a world and, in turn, affecting that world. The making of images is also always the making of meanings. And neither images nor meanings are fixed. Both involve the participation of the viewer in interpretation, and both can change over time. Images, like knowledge, are always partial, situated, and power laden, with unpredictable impacts and forces. As a part of ethnographic research practice, photographs have affects themselves, moving back and forth between site, researcher, participants, surroundings, and memories.

How do we really see these familiar terrains?

The temporal arc of a photograph runs from the instigation to make the image, to the material technologies and embodied practices that shape how that image is captured, stored, viewed, and reproduced. The persistence of the image continues to make worlds beyond the present in which the image itself was made. Each of these twists and turns in the lives of images becomes integral to the ways meanings are made from them.

Things matter not because of how they are represented but because they have qualities, rhythms, forces and movements.[4]

SARAH / YARD DIARY / 23 AUGUST
How do you feel about your yard today? OR Write down any additional comments or thoughts below.
My tomatoes are splitting on the vine – my sister says it's from inconsistent watering – which I believe because water from rain has been in short supply this summer and water from a hose is never as good as rain.

LORRAINE / YARD DIARY / 31 JULY
What is the weather like today?
Hot and humid – very uncomfortable for me

How do you feel about your yard today? OR Write down any additional comments or thoughts below.
happy to see boulevard weeded
glad to have a shady place to sit on a hot day
still hoping for rain! watering is NOT the same

LORRAINE / YARD DIARY / 4 AUGUST
Did you spend any time outdoors around your home today? Yes
If yes, about roughly how long, in total? 40 min
If yes, what did you do?
got paper
went out through front yard to go to theater – came home through yard late
sat out and read paper while kitty explored
went through front yard again to go to son's house up the street from my home
What is the weather like today?
Cool! sunny – rained 1.75" last night
*How do you feel about your yard today? OR Write down any additional
comments or thoughts below.*
Glad it is watered so I don't have to do it today!

SARAH / YARD DIARY / 22 AUGUST
What is the weather like today?
It threatened rain all day and then around 3PM it rained lightly. I opened the
porch doors so I could see the yard wet and glistening – the color becomes
so much richer in the rain.

THIS IS A MASON BEE HOUSE. And it looks like they're dead or something. The mason bees bury their eggs in there in the mud, and then the mother/queen flies off and then they just emerge through the mud. When they were starting to have problems with honey bees, then mason bees were what gardeners and people were using in Oregon, so my brother in law built that house. I asked him about it and he built the house for me; it's supposed to match our house. He said, if you see them in your yard, then you have them, and so if you put the house there, you'll help them – and I did go out in spring and see them around that tree and see them all over those blossoms, so I know we had them, and he said all the states have them, you just have to encourage them to come.

THE MILKWEED just kind of took over the front garden—I guess the garden kind of changed purposes. I let the milkweed come because I was collecting the monarch eggs, and last year, it was more than a hundred. Well, the past two or three years has been more like 200, and this year it'll be a little less. I was leaving the milkweed to grow so I could collect the eggs off them, and then I was taking all the leaves off to feed them. And then I ran out of food! So I went all around the neighborhood and found all the milkweed patches, and met a neighbor, and her whole yard is just milkweed.

I was feeding twenty-five at a time, it took – you know, when the caterpillars are big, they go through about a leaf a day, when it's not on a full plant. So I'd go in the morning and afternoon.

Most people, they wouldn't allow milkweed. So milkweed kind of took over this area, but then I use it for a different purpose. It changed a little bit. Some people like the more manicured look and those are the people that kind of grumble about people who let their gardens go [me].

IN THE SUMMERS I PLAY, I recharge my batteries. I catch up on reading for fun, just being outside, playing with the dog. In the summer I'm relaxed and enjoying my neighbours and home.

I enjoy eating together the most on the deck in back. Eating together is our time, family time. It is a time to talk about the day, to plan the next day or week or what is coming up, we get to be together. It's uninterrupted – we don't answer the phone, we don't get the door.

The back is fenced, so we don't have to manage the dog much, and we spend time there because it's shady in the late afternoon, which is nice in summer.

I always plant the impatiens around the tree on Memorial Day weekend.

Tina plays catch with her dog, Hoss, toward the end of our yard visit. She lives with her husband and two cats in South Minneapolis and is a kindergarten teacher.

Minneapolis has a population of about 240,000 cats and dogs, and a human population of about 380,000. Of those pets, about 110,000 are dogs. Of those 110,000, 8 per cent are licensed with the City of Minneapolis Animal Care and Control Department. Licences are required by city ordinance for all domestic dogs, cats, rabbits, and ferrets.

Plants

People devour, anticipate, fret over, destroy, cut, break, snap, mow, water, share, move, compost, haul, dig, carry, sow, wait for, protect, divide, feed, trim, allow, scratch, sift, mulch, pull.

Animals

People trap, feed, walk, release, scratch, spoil, replace, fence in, fence out, curse, house, observe, listen for, watch, notice, chase, herd, stroke, throw to, leash, unleash, kiss, treat, swat, endure.

Instead of thinking of organisms as tangled in relations, we should regard every living thing as itself an entanglement.

Thus, far from inhabiting a sealed ground furnished with objects, the animal lives and breathes in a world of earth and sky – where to perceive is to align one's movements in counterpoint to the modulation of day and night, sunlight and shade, wind and weather.[5]

WHAT I ENJOY THE MOST is watching it grow. I just love watching it grow. And I love tending to it, like keeping it clean – but I like to just sit here and look at the greens. See the color and the leaves. Such a beautiful green color, with the veins and all that. I just love to look at it.

Get up close to those greens. Remember I told you about how beautiful the color is? I just love the color. And over here, these leaves are getting *big*. The leaves, they used to be pretty big – but now. *wow.* Take one of these big leaves to Chicago and let my sister see it! Word about my garden is all over. All over. The word has been around, about that guy with that garden on the corner.

Morris's yard in North Minneapolis was one of the only yards in this research that also was a site for commercial activity. Morris has been gardening large areas of his back and side yards since 1989, expanding more and more. After two or three years of giving away his greens, he has been selling bunches of greens for $1.00 to neighbours, friends, and passers-by for the past twenty or so years.

MORRIS: About three to five years ago, a woman was really insistent about me *raising* my price. You know, you go to the grocery store? It's around a dollar. So I don't want to *outdo* the grocery store, so high that people don't wanna buy 'em. They want to eat them. If I got enough, I'm satisfied with it all.

If I keep at a dollar, I don't have to go through all that, making little change. But looking at it now, with all the expenses, you know, the water and feeding the plants. I *may* go to a dollar and a quarter. But like I said, that wasn't my plan. And you know, that would help me out as far as the expansion out here [planned for next season]. But then I gotta give people change, little change, coins and all that.

URSULA: Do you keep track? Do you keep track in your mind, how much you are making or how much you've spent?

MORRIS: No. *No.* But I don't touch it. I don't touch what I make! I put it aside and I don't touch it. I will not touch it during the selling season. There's been times I didn't touch it for two years and I didn't even know what I had. I won't count it, separate it, I don't count it – what I make. That's *irritating*, too. Every time you sell it, keeping track and all that – ugh.

I DON'T WANT TO GO back to all aspects of the 1940s [victory gardens] but this is about having more of a purpose, more of an intention about raising vegetables and having a responsibility for the future. You know, everyone has some kind of vision of how it might or could be, but nobody really has much time to make it happen. We are trying, and trying to be intentional about it.

JOE / YARD DIARY / 13 AUGUST

How do you feel about your yard today? OR Write down any additional comments or thoughts below.

Cleaning the fountain is a fairly quick matter of using a jet of water to flush out the greenish water and algae on rim, and then use a slow flow to refill the basins. I replenish the algaecide to finish (shake a few drops from the bottle).

JOE / YARD DIARY / 15 AUGUST

Did you spend any time outdoors around your home today? Yes
If yes, about roughly how long, in total? 1 ½ hrs
If yes, what did you do?
1 hr – reading on the porch
30 min – having dinner on porch

Did you see anybody while outdoors around your home today? Yes
What was your interaction like? Dinner with my wife
How do you feel about your yard today? OR Write down any additional comments or thoughts below.
Regret that we haven't had many dinners on the porch this year because of too hot weather and TV habits (we've started to watch the evening news and eat while watching).

JOE / YARD DIARY / 19 AUGUST
Did you spend any time outdoors around your home today? No
How do you feel about your yard today? OR Write down any additional comments or thoughts below.
We had two political signs in the front yard, and when we got back from a play (afternoon) and dinner at 8:00PM, we saw that they had been taken – we felt angry (me) and discouraged (Helen) that such a thing could happen in our neighborhood. Now, we have doubts about replacing the signs.

IT WAS SO MUCH FUN. And laughter, and the lilacs. There was a lilac tree right here, I think. It was really the neighbor's tree, but it was all over here. And you could smell lilacs in the spring.

Betty grew up in Chicago, after earlier generations moved north during the Great Migration.

We grew up with a yard, but not so much *garden*. We played in the *yard*, that's where we used to be, that was our home, outside in the summer. And the winter – we didn't do a lot. We never were a family that went out skating, or in the snow. It was like, just get in! But in the summer, when summer came, then the yard was *ours*. And just doing different things, just building stuff, making stuff.

Trees are the measure of things. A tree grows, and we measure ourselves against it.[6]

WE'VE SEEN A COOPER'S HAWK, and red-tail hawks come through. One night I came home and there was a barn owl sitting back here, which I didn't realize are in the city. You know, the barn owl was right there. But – they were probably eating the voles and then the little black mice, with the short tail. And when they get under my deck, and that's when the house kind of smells. Supposedly they don't get in the house, but there's also some mice that I catch occasionally.

Well, actually I've got five traps now – so what I've done is covered up the hole, and I'm trying to see if there's some activity. Sometimes I put peanut butter on there, and then the ants come and eat the peanut butter.

I handed out flyers and met Adrienne; we chatted on the front stoop and made plans for a proper yard visit soon. She insisted I make the round trip on my own that day, through the south side, back, north side, and front yard spaces. "It's kind of a sequence, really." She told me as people stop by more and more, curious about the elaborate fountains, fairy scenes, and sculptures, if they are home and able to come out to say hello, she and her partner and kids often invite people up to make this circuit.

There were interesting stops along the way, a little bench in an alcove, fish ponds – the main feature of the backyard. I looked up toward the back of the house to find an elaborate deck system with at least two levels and a shaded screen tent on top. And then as I walked back out into the front yard along a stepping-stone path, there are so many little houses and statues and scenes. Adrienne told me later that she works with survivors of abuse, and the small figurines and miniature landscapes sometimes spark conversation and healing.

Making Photographs

Although I devoted significant time to developing several distinct photographic perspectives in the project, the one that mattered most, and from which these images are largely drawn, is tuned to feeling, gesture, and moment. Part trace, these images document our movements and the features of each yard. Part perspective on registers circulating within our conversation. Part encounters between bodies in the space of the framed image, the framed yard visit, the framed yard. Part atmosphere captured.

"Show me your yard."

These are images made in the heat of the moment, juggling notebook, voice recorder, pencil, and conversation. Most of these photographs get the details down, record our movements in the yard, capture what people show me and in what order. Something pointed out to me, something they think worthy of saving in an image. A figure made against ordinary ground. A slight pause while I photograph something about which we were talking, participants moving aside foliage to better show me and the camera something important. A close-up of hand and plant touching. Then a scene caught mid-range. Sometimes furtively framed to include inhabitants trying to move out of frame. Some people were reluctant to be photographed, darting out of the frame as soon as I raised the camera. Some proceeded with their activities, seemingly unbothered by the camera. Sometimes together we ignored the photographs I made.

Performing research through shifting relations and intersecting identities. Guest, host. Visitor, visited. Person of colour, white person. Old woman, young woman. Gardener, non-gardener. Neighbour, friend.

Lorraine sits in one of her favourite backyard spots, an outdoor bench with cushions. To her left is a mosaic stone her parishioners gave her when her husband passed away.

She moved it with her to this yard from her previous yard.

THE PREVIOUS OWNER of this house told me – they were both musicians, they didn't garden – he said, if you ever take that tree down, don't tell me. Cause they planted that when they had a miscarriage. Probably twenty-seven years ago now. So. It's a pretty nice tree, and so far, I've had it trimmed.

I REALLY LIKE that the kids can just be out here – and I can be in
the kitchen. That was one of the little things at our old house –
that I couldn't keep an eye on them and work in the kitchen at
the same time. I can be moving around here just to check on
them every so often.

Juliana is a doctor who lives with her husband and two children under five.
Her husband works part time and from home. They are relative newcomers
to the block, moving to this part of South Minneapolis earlier that year –
although when I asked about this, her identity as a "newcomer," it didn't
resonate strongly with her. I thought it might after talking with residents
who have lived on the block for twenty years or more.

OUT HERE IS *pure symbiosis* between house and yard, hand and glove, part and parcel, if you will.

I drive around on my bike, I look at houses and yards. I've always had a thing where – why don't people build *fences*? There are just all these yards.

Every yard should have their own fence. They should be wherever they should be in each yard. Every yard is different. Therefore, every fence should be different. And they should have gates on them. To me, all these expanses are just so incomplete! Why don't they have fences? They're so *easy* to make! Like this – these are just two by fours. I got all this wood free, that guy next door was just throwing it out. I said, I'll take it. I whacked it down, and I made these – you know, I took my saw, and made a little round thing on top. It's all free! People throw this stuff out.

I think just nobody has *time*. But there are people that do have the designer fences – they're crooked! Look at the top of this fence, it's real level. I'm picky about that kind of stuff. There's ugly fences and there's good-looking fences.

To me, it's like – *complete* your house. It's like the house just stops. There's no intimacy, there's no effort, there's no entry, nothing you move through.

It should continue to the sidewalk, you know? Invite me in.

It changes everything. But. Well. Fences. A lot of people that have houses, they can barely make their mortgage payment the way things are now. So.

John showed me in great detail the fence he had just built, including a front gate requiring careful attention to level and swing. The slats to the right and left of the front gate, pictured, were modelled roughly on proportions from Chartres Cathedral.

Yard makes *ground* to the *figure* of house. A yard can be *figure* made of intention or neglect, other yards defining its edges, its *ground*. A yard can be *figure* made of other figures, accumulations over time. Sometimes a yard is simply *ground*, as in earth.

ANN: We have sort of divided green space. Bonnie's backyard is [pause] *Bonnie's backyard*. I started sneaking hostas in the front. I thought if I just put in a few, she wouldn't notice. Every once in a while, I'd dig up another row of grass and put some plants in. Bonnie's very observant but not necessarily apt to comment.

BONNIE: The grass didn't do well out there because of the big trees, once the trees got big.

ANN: So I just kept encroaching on the grass. And you would be pleasantly surprised from time to time. Is this sounding familiar? [laughing]

BONNIE: Yeah. But we don't *sit* in the front yard much. We occasionally do.

ANN: When there are drug deals on the block. The block club sends out emails periodically, and then, for a period of time, a lot more people are out in their front yards.

BONNIE: It's been fairly consistent. It's amazing. That if you know what you're looking for, you can see it happening … I think the drug dealers have learned to come to the good neighborhoods, as opposed to – they really don't cause any trouble.

ANN: It's not like they're knocking on the door.

BONNIE: It's usually affluent, young kids, because you know they don't have a job good enough to support that kind of car.

To perceive the environment is not to look back on the things to be found in it, or to discern their congealed shapes and layouts, but to join with them in the material flows and movements contributing to their – and our – ongoing formation.[7]

Sometimes perceiving environment does include looking back on things. Sometimes that is the thing.

THE THING that's so interesting about this perspective, is that you really get an appreciation for all of these textures. And when the sun changes, the colors change, and it's always changing. So, from *this* perspective, *that's* the thing.

That over there is more sculptural, and you see the different designs of the trees. That is a whole different experience – laying on that hammock. Oh my gosh! And then this is not a bad place to sit back here, to sketch, because you can see – well, I don't know, it's about similar.

I LOVE THIS PORCH. It's just a three season, but it's nice. Because it's so giving. These are the viburnum bushes, I have to keep them trimmed cause I don't want them to come too high.

I come out in the front in the night. Cause I don't wanna be seen – cause I have found out, people stop too much when I'm out here. And I don't want to be, you know, always disturbed. So I'll come out here, I find myself sitting out here in the nighttime more. But I still like the back. That pergola area, I just love that area. And I like to just sit there, and people come by and I can see things.

Like the environment of which it forms a part, the building neither encloses the inhabitant, nor is it disclosed from within.[8]

We took turns standing on a crate to peer over the top of the old fence. Beyond our perch stretched a sea of head-high plants, shrubs, milkweed. An old car planted at the far edge of the yard.

When I think of a landscape I am thinking of a time.[9]

I WOULD SAY our yard is kind of our pseudo up north oasis.

It just doesn't matter what's going on, all around you or even in your head, that you can come out here and it just really kind of washes away. I don't know what that would be. But, it's like taking a small, little – and a small little vacation at home. Everyday! Whenever you decide to use it.

Those are beautiful trees. You can go over there and it smells just like up north. You just feel like you're someplace else. When it rains, you can smell that piney scent.

I WISH OUR HOUSE was more connected to the yard – there's one window on the landing of the stairs that looks out to the side and I sometimes stop just to look out whenever I can.

As the life of the inhabitants overflows into gardens and streets, fields and forests, so the world pours into the building, giving rise to characteristic echoes of reverberation and patterns of light and shade.

It is in these flows and counter-flows, winding through or amidst without beginning or end, and not as connected entities bounded either from within or without, that living beings are instantiated in the world.[10]

SHERI: It's really very unusual that we would do something like this [build the patio].

Well, I have these ideas that I want to do. I have a bunch of those big rocks, so I thought the rocks could be just underneath the deck and then maybe some of the leftover gravel up under there. And then I'd love to get a little fountain that makes noise.

We looked up online how to do it, and we called up my step-dad and asked him for ideas, and so it really isn't a hard, hard thing to do. So that's good.

JIM: It feels good to be doing this. It feels good.

Husband and wife, Jim and Sheri, talk over the side fence with a hobo visitor to their neighbours' yard. These are neighbours who are also self-described hoboes, regularly riding the nearby rails, and hosting many visitors in their yard and home who come from the tracks. Although I was unable to conduct a full yard visit, I talked and visited with Fanny, the matriarch of the household, and Hobo Queen multiple years running – crowned at the annual Hobo Convention in Iowa. Fanny is an accomplished gardener, with a special interest in cultivating native plants for butterflies and other insects.

Jim fills in gaps between patio bricks. Making a place to sit.

When I met with them the following year, Sheri told me that they used the patio a few times – "We had some people over once, and made a fire. In the fall, when it was cooler. It was nice."

TIM / YARD DIARY / 2 JUNE

Did you spend any time outdoors around your home? Yes
If yes, about roughly how long, in total? 7 hrs
If yes, what did you do?
returned garbage container, napped on deck, filled bird feeder, weeded garden, napped against a tree, stacked rocks in rock garden
How do you feel about your yard today? OR Write down any additional comments or thoughts below.
Satisfied at day's end with what I accomplished. Have rock garden 97% weed free, hope to finish tomorrow.

TIM / YARD DIARY / 3 JUNE

Did you spend any time outdoors around your home? Yes
If yes, about roughly how long, in total? 4 hrs
If yes, what did you do?
weeded rock garden, did some rock placement, weeded around back of house, rested in shade of ash tree, watched birds at bird feeder, herded rabbit into neighbor's yard, planted tomato in backyard, watered tomato plants and rose bush, smoking cigar and sipping scotch in garage while recording this, listening to baseball game.

TIM / YARD DIARY / 4 JUNE
Did you see anybody while outdoors around your home today? Yes
Who were they and what did you talk about?
Mary, neighbor to north, she looked out window and we exchanged greeting.
How do you feel about your yard today? OR Write down any additional comments or thoughts below.
Tired. Not satisfied. I tried to arrange the marigolds into a peace symbol but don't think people will recognize it as such. Developed a blister on my hand that opened.

TIM / YARD DIARY / 6 JUNE
What is the weather like today?
sunny 80s, then rain around 6:30
How do you feel about your yard today? OR Write down any additional comments or thoughts below.
Our lawn service mowed the grass and yard always looks sharp afterward
Relaxing. enjoy placing and positioning rocks – therapeutic, Zen?

TIM / YARD DIARY / 9 JUNE
Did you spend any time outdoors around your home today? Yes
If yes, about roughly how long, in total? 2 hrs
If yes, what did you do?
In the morning did a little weeding, but it was too hot to continue.
In the evening my wife and I enjoyed a beverage on our deck.
Did you see anybody while outdoors around your home today? No
How do you feel about your yard today? OR Write down any additional comments or thoughts below.
In the evening once it cooled down, we were chatting on the deck and my wife commented that she would miss our yard if/when we move away. We have been talking about someday moving once we are no longer able to maintain the yard/house ourselves.

TIM / YARD DIARY / 11 JUNE
Did you spend any time outdoors around your home? Yes
If yes, about roughly how long, in total? 2 1/2 hrs
If yes, what did you do?
read a John Grisham novel in the garage

I WANTED to pull my car up to a backyard instead of into this driveway that goes on forever!

Sarah told me about the previous owner, a longtime resident of Northeast Minneapolis, and his pride at the "one-hundred-foot driveway." She chuckled and groaned, but also appreciated that, for him, this was likely a sign he had really arrived.

She and her partner built large planter boxes on wheels that can be reconfigured, though the boxes stay in the same place most of the time. In high summer and early fall, plants, many vegetables, almost entirely obscure views in and out of the yard, and form a zone around the door to the back porch.

Afterlives of Images

Sifting and assembling image and text makes meanings on the page. In the abstract, images have the capacity to document enormous numbers of details, and structure parts of yard visits. These images also set up the conditions for how knowledge about yards is produced later, when I am no longer right there in the place or time. Writing with images activates senses and temporalities over the course of a long project. Images begin to stand in for moments, inviting affective memories. These openings back into an encounter disrupt linear temporalities of analysis and representation, and of researcher and writer.

Similar to the intensively experienced aural atmosphere of transcribing yard visits, these images interrupt, draw out, and contradict my own memories. Small details such as the condition of paving, irrigation systems slightly obscured from view, or the contents of neighbouring yards reveal habits of upkeep and the resources available for such. Searching for affirmations of memories, chasing the right words, remembering how a participant moved in their yard.

Summer air on summer skin, summer sounds in summer air. Images elicit sensory experiences transported through time and space. An image takes me to Sandra's yard. Now I am there, helping load up a wagon with weathered frog figurines, wheeling its wobbly heft out to the front yard. Or we sit together under the pergola, on a decorated outdoor sofa with pillows, sheer fabric in the breeze, appreciating our carefully cultivated surrounds.

What are the possibilities and limitations of photographs to better understand this affective force?

Through stories we make the world. In this chapter, photographic vantages together with words – as partial and laden as they both are – make stories. What is affirmed in the telling of particular stories, and what is obscured, left out of frame? These views entangle readers, writer, inhabitants, yards. Through these connections to others, could we make other possible worlds?

It takes work, but property settles into a stable and bounded concept, day by day, block by block, parcel by parcel. Yard by yard. Plants and animals become drawn in, too. Owned and regulated, yards are designed, and yards are maintained.

Morning coffee on the back porch in Northeast Minneapolis.

Sandra's neighbours and friends, the "Alley Cats," as she calls them, in the garage across the alley. These three generations of fathers and sons, and friends, regularly help her with tasks around her place, as well as spend time in her yard – with her, or on their own. For instance, they regularly sit around her large outdoor fireplace, a spot that comfortably seats fifteen to twenty people.

4

PRACTISING PROPERTY AND LIFE IN COMMON

There may be no more potent trope of private property in North America than the single-family house, surrounded by a moat of uniform lawn. But this image obscures the surprising variation in the forms, meanings, and practices of property in yards. As sites of social and ecological differences, yards add up to something more than the sum of their parts. They can be meaningful landscapes beyond private property boundaries and logics. Indeed, yards function for some of the people who live with them as commons, specifically as shared territories and as sets of shared practices. Yards can be understood as both commons and private property. What might the yard tell us about the nature of property, especially as people live with the same yards over time? How do yards and yard practices disrupt and reinforce private property and new forms of commons and commoning practices? I take a relational approach to thinking about yards, which acknowledges and explores their dual nature as, on the one hand, a bounded territory and, on the other, always also a set of relations. Yards are parcels made from survey markers, a potent distillation of logics of private property. And, at the same time, yards are shared connective tissues spanning neighbourhood boundaries, watersheds, forest canopies, and property bounds.

This dual nature of yard connections and individual ownership reverberates through environmental policies and projects geared toward individual adaptations and environmental care as best management practices. Municipal environmental strategies increasingly see yards as part of open space for vegetation and stormwater management, linking them up with geographies beyond individual lots, while also encouraging individual owners to adapt behaviours, make rational choices, and deploy best gardening practices. Through these shifts toward more sustainable practices, environmental managers at various geographic scales attempt to shift responsibilities, risks, and potentially benefits onto individual private property owners. Some of these shifts might be aspects of management previously under the umbrella of municipalities, but other shifts may actually be emerging with new thinking about the role of green spaces within urban environments.

Examining the varied ways people cultivate boundaries and borders lets us understand the role of privately owned green spaces within landscapes of property values and broader urban geographies, and the ways environmental managers understand these individually owned private yards. Yards as private property embody many of the dominant discourses about bounded territories, individual rights, and concerns about property values. But I also explore the ways we might consider yards to be one form of urban commons. Informally shared yards highlight how inhabitants maintain and experience heterogeneous urban commons, and the ways yards participate in commons as shared territories. Yards facilitate commoning practices by providing places for shared practices and meanings to emerge through plant sharing and exchange. Neighbours routinely share and exchange particular varieties of plants – planting and cultivating commons distributed across property distinctions and made possible through the plant tissues themselves. Across these cases, surprising diversity emerges in the forging and cultivation of common experience, the roles of plants and gardens within these experiences, and the rhythms of cultivating and inhabiting the common life of yards. Throughout, these commoning practices are intertwined with logics of private property ownership.

The Natures of Property

Property is, at its core, a practised set of social relations. People practise property. The practice of property involves material and immaterial entities: owners, financial infrastructures, state regulations and enforcement, as well as land, territory, and negotiations of difference. Property happens within multiple temporalities. Property can be a contradictory and, at times, shifting terrain of individual and collective interests and practices.[1] The set of relations that constitute property become territorialized – in other words, given shape as bounded parcels of land – and defined in terms of private, individual owners. This process of territorialization requires effort through practical infrastructures such as legal categorization, municipal governance, financial systems, as well as shared meanings to enforce and constantly shape these bounded and landed territories. Through all of this ongoing effort, private property appears to be constituted of stable, fixed, and neutral territories. As we know from the work of feminist geographers such as J.K. Gibson-Graham and others, the making and practice of property occurs through many diverse economic practices.[2] A focus on dominant capitalist categories of work, leisure, home, time, and value can obscure the ongoing making, meaning, and maintenance of property. In short, property organizes the world for and with us, and not just in relations between people and things – but between people and people.

Private Yards

Certainly most yards fall fairly easily into private ownership models of property. People are concerned with maintaining their financial investments – for most homeowners, one of the most significant of their lifetime. Yards are spaces that can communicate capacities for neighbours to maintain certain standards – some enforced through regulation but usually more pressingly through the expectations of neighbours. Along with this sense of individual responsibilities, there are also individualized rights that people call on to make sense of aesthetic interventions or gardening practices in yard spaces. Within the logics of private ownership, participants more or less consciously defied these expectations.

Nils cultivates his tidy corner lot in Northeast Minneapolis, across from a community garden and small field between his house and the railroad tracks. He told me about the housing crisis and his now upside-down mortgage in a matter of fact way. "This is my first house, I bought it in 2006, which was the worst time to buy a house, at the peak of prices. I also bought it with a non-prime mortgage like most of the people then; it's not a good one. It's not a good situation but at least I can manage." Before this house, Nils rented in other areas of Minneapolis and while a student in Sweden. As we stand in his meticulously maintained yard, Nils tells me about his earlier efforts at gardening around a small patio. "I used to cut the grass with my scissors; it was a small area, about five feet by ten feet."

When I ask him what it feels like to own this place, he tells me it's different than renting. "When you own the land, it makes a big difference for what you feel like doing. When you just rent, you don't really know how long you'll stay, so you're never really sure if you're gonna get the benefit out of your effort. When you own, that is pretty clear." Nils also connects ownership to neighbourhood-wide challenges of empty foreclosed homes and the inability to maintain much beyond simple mowing, and he articulates yards as a kind of barometer revealing the capacities for individuals to keep up with demands of work, mortgages, and other obligations. Having gone through the drop in house values, he seems to be sympathetic to these challenges, and aware that, with his salary as a researcher at a university, he's able to keep up with payments. In addition, his farm background enables him to have what he describes as "pretty obvious" knowledge: "I can see when things need to be done, see when plants are overgrown, and when they need to be watered."

A few blocks away, Sarah and I walk through her small front yard. She stoops to pull weeds as she talks, "Like everyone says, 'Your house isn't worth anything anymore!' And I'm like yeah, but it's worth something to me. It's my shelter, it's my place where I work, it's the place where I eat, it's where my dogs live, it's where my partner lives." In 2011 and 2012, residents in Northeast Minneapolis were just beginning to find their feet after the housing crash, and it was a topic of common concern across households. Sarah's take on it was not uncommon, as people articulated some distance from a fixation on property value. She continued, "I don't want to think about its value all the time in some arbitrary market. I mean it truly is just arbitrary and it depends on somebody else, you know? You decide to choose which systems are meaningful."

Environmental planners, managers, and advocates mobilize motivations based on individual concerns with property values to try to shift individual behaviours toward best management practices. But this mobilization takes shape within an interesting contradiction. Often it is through the very imaginaries of urban environments, such as watersheds, foodsheds, forests, and wildlife habitats, that cannot be contained within the bounds of private property or political boundaries. Even as environmental managers and advocates reimagine urban territories through linkages across conventional boundaries, such as forests, watersheds, and foodsheds, the individual private homeowner often remains the locus of possibilities for adaptations in response to environmental concerns. These individual owners are understood as rational decision-makers and consumers. Arguments in this vein point toward the benefits of best management practices in yards as enhancing property values. This belief can feed into intensifying processes of green gentrification, often without reflection on the hazards or social impacts of these material practices or changed individual behaviours.

But the private property aspect of yards does not tell the whole story. Yards are not only private, individualized realms. Pulsing throughout these interior-oriented experiences and practices, worries, and pleasures, are social and affective relations to do with world-making. These are relations within affective ecologies of care, joy, engagement with the more than human. There is also failure and frustration felt in the bones, over and over through everyday practice – these are never only individual, private, interior, and are always in relation to bodies, surroundings, materials. Can we see more than individual bourgeois desires and reflections in yards? I say yes. Within experiences of yards there are critiques of private property, and certain possibilities for other forms of social relations through property. One way to find these is to look for urban commons within these connective tissues. They are not always there, but the disruptions we can find are worth further exploration.

Yards in Common

Paying attention to commoning practices allows us to build a more variegated understanding of urban commons. The traditional meaning of commons as collectively owned and managed natural resources, the primary purpose of which was to provision raw materials for subsistence and livelihoods, is giving way to a broader meaning that encompasses views from natural resource management and ecology, development studies, political ecology, and economics.[3] From their original rural contexts, research has focused on traditional forms of commons that support livelihoods such as common-property and common-pool resources within urban contexts.[4] These commons are biophysical resources such as water, waste, or urban forests. Conceptualizations of these new commons share a more expansive view of how common resources might be constituted, emphasizing that these resources are not only material, but also social and cultural – for example, commonly held scientific knowledge and intellectual property, or even shared culture itself. The production and handling of meaning through people's more mundane lived experiences remains largely unexplored and unnoticed.[5]

Beyond both these material and immaterial imaginaries, commons have come to resonate politically as a means to resist enclosure and privatization. Life in common – in material spaces and through everyday practices – holds within it the potential of social relationships forged beyond dominant narrow market logics. Furthermore, this sense of commons as political potential is built around observation and analysis of diverse sets of practices, through which potentially transformative socialities may be produced in conjunction with shared resources.[6] The transformative power here, according to such recuperated interpretations of commons,[7] arises out of commoners producing shared resources, and their recognizing previously unseen means of provisioning human needs together, on different terms than the logics of capital.

In addition to this focus on varying forms of urban commons, multiple temporalities constitute important dimensions of commons. While the question of long-term sustainability has often been posed about commons, more invisible everyday temporalities may also be key to the making and maintenance of urban commons. Daily life is full of routine repetitions, perpetuating sedimentation and the reproduction of certain social and political relations. At the same time, the potential for difference within these habitual repetitions animates calls for alternative and emergent modes of collective life.[8] Temporalities of regulated time, integral to and in the service of modern linear systems such as capitalist production, interfere with cyclical rhythms as visceral, vital, and lived in excess of narrowly imagined economic terms.[9] These dimensions of rhythms constantly interact, with serious social, material, and affective consequences.

Just as temporalities are multiple and entwined, the logics of commons and commodities need not necessarily be distinct or mutually exclusive, and may even be dependent on one another in unexpected ways.[10] How, then, might private property participate in urban commons? Privately owned yards are inherently interstitial. For example, in spatial terms alone, they are situated between homes and also between home and street. This in-between nature of yards offers rich opportunities to further understand and complicate distinctions between private and public property in lived space,[11] as well as the variety of practices that constitutes everyday domestic and neighbourhood life.

Research with urban inhabitants living with yards allows us to expand our understanding of urban commons beyond notions of public territory or immaterial and abstract socio-political connections. Commoning practices are a vital aspect of how people produce, cultivate, and adapt the complex and diverse urban worlds that yards constitute. Yards and the everyday practices that take place in and through them constitute one kind of nodal point in the making of urban commons. Here, logics of private property and logics of commoning become interwoven.

Front-yard Commons

Commons are always in conversation with their surroundings. Tim's front yard is a thick jumble of plants, reused metal fencing and supports, and wire sculptures. Vegetables grow in and among the waist-high tangle. Walking north along the block, you feel the difference of moving from the more prevalent grassy front lawns and boulevards, through two shaggy green edges that brush past your legs. Through Tim's striking front yard circulate social relations with his neighbours. He and six or so neighbours on his block have cultivated an informal community garden here. This

collective use of front-yard space produces food, as well as less tangible but meaningful encounters and habits.

Contrary to many urban gardening projects, Tim's front-yard garden emerged as an informal community space with little forethought or intentional planning. He's nonchalant when describing how the front yard community garden emerged:

> Five or six years ago, I planted squash and pumpkins in the front yard, as a way to try to limit the mowing. And then I guess it was four years ago, when I was looking out there in the springtime and it was maybe April, and a couple of my neighbors said, "If you're gonna plant this, why don't we make this a community garden idea?" And then not even an hour later we had six people with shovels digging the whole thing up. Dig it up!

The garden has no formal or individual plots – it's just one garden, and people take what's available when they want it. Tim says, "It's not like a lot of community gardens are, where everybody gets your own fifteen square feet. It's more like, what do we want to plant, what do we want to grow, and where is the best place for it to grow?" Basic decisions about what to plant happen all together early in spring, and then the six or seven other gardeners bring seedlings and plants. One involved neighbour works at a nursery and has significant gardening experience. Another built fencing that serves as a trellis system. The tasks of cultivation and maintenance over time are generally loosely shared. Tim pays for water, and so the others involved provide more seeds and seedlings, as well as materials such as tomato cages.

The ways Tim practises his property are important to his own sense of self and how he positions himself in terms of wider social expectations. He jokes about being the "baron landholder," with his neighbours as serfs working the land. His approach to his property is unconventional and relatively open to what might emerge. "When I say to people, 'Well, why wouldn't you turn it into a community space?' they say – 'Why do that? You can do that?!' and I'm like, sure, why not – property ownership is a bit fleeting anyhow, and many cultures don't have that same view of property anyway." Later, he tells me he thinks he just has "more tolerance for people just wandering in, willy nilly."

Like many decisions about planting vegetables in fairly dense urban areas, enough sunlight and accessible terrain can determine where and how gardens are made and maintained. When I asked how the community garden got started, Tim explained it in terms of natural resource constraints along the block. He said, "My yard is the only one with late afternoon sun. The sun is so intense, it just bakes the house. On this side of the street, everyone north of me has trees, everybody south of me has a hill, so I'm really the only one with the ability to do this." Neighbours pointed to his garden as an interesting feature of the neighbourhood, something about which they seemed not just tolerant, but positive.

Tim's account of his front-yard garden offers a self-effacing and action-oriented reimagining of the possibilities of what a front yard might be – materially and socially. Beyond the sheer material features of the front yard, the neighbours clearly have a strong sense of satisfaction and enjoyment with one another. Tim's front yard provides a space and time for them to work together, respond to biophysical conditions along the block and throughout the changing seasons, and cultivate an unusual use of a front yard – important to a sense of self and meaning in place.

Cultivating Common Practices

Collectively cultivated commons are shared territories and enable common experiences. The everyday practices and labours of plant sharing and trading forge commons in and among individual yards through the tissues and biophysical capacities of plants themselves. Certain perennial varieties persist and thrive in a range of environments (and without much human involvement), which enables especially working-class areas of the city to cultivate and foster a particular kind of common life among yards. Exchange is seen as meaningful here as a way to provision yards with plants considered appropriate in terms of aesthetic taste and biophysical climate, and as a recognition of homeownership and inhabitation. These practices of plant exchange and sharing also can be understood as an exchange of socially meaningful matter, responsive to common cultivation, seasonal rhythms, and skilful experience. Such exchanges provision meaning within the neighbourhood landscape and open up possibilities for relationships between more and less experienced and skilled gardeners.

Tiger lilies and hostas are the two varieties of plants that came up again and again, across all neighbourhoods. Both are known for their ability to thrive under resource constraints. Tiger lilies, or orange day lilies, are almost indestructible, able to grow and thrive even in very dry, sunny, and hot conditions. Hostas thrive in shady areas, under heavy tree or shrub canopies. Most participants who spoke of them implied that these perennials are necessary for challenging areas of the yard, or good "starter" plants for those less interested or experienced at gardening. With a few exceptions, residents across different neighbourhoods and with different degrees of gardening skill expressed a kind of obligatory acceptance of these plants' sturdiness and necessity, as well as their role in social exchanges between neighbours and friends.

Some participants also articulated class, history, and generational changes in garden ideals in their understanding of these particular plants. For Dan, a long-time resident of Northeast Minneapolis, tiger lilies and hostas are shot through with markers of class and neighbourhood identity. Historically, this part of Minneapolis is a white working-class area in which major industries such as railroad yards and manufacturing are situated and continue to shape the physical and social dynamics of the area. He said,

"Well, I'm pretty blue collar about my yard – yep, it's tiger lilies and hostas for me." Dan expresses both pride in the neighbourhood and a sense of how outsiders might see plant choices in the neighbourhood, nested within an awareness of broader urban geographies.

Becoming Established: New Homeownership and Circulating Gardening Skills

Although perennial plant sharing is practised throughout all three study neighbourhoods, there is a stronger sense of commitment and articulation of this kind of exchange as a philosophy in the two neighbourhoods that are more working class: Northeast Minneapolis and North Minneapolis. The most common refrain I heard from both long-time residents and new homeowners in these areas is that practices of sharing perennial plants can be a means to help newcomers get established in their yards precisely at the moment when they might have little extra cash or gardening knowledge. In Northeast Minneapolis, Jack, a relative newcomer, has been actively trying to trade perennials to replace the tiger lilies and hostas he inherited when he bought his first house eight years ago. He sees nearby yards as sites for swapping and sharing plants, as well as a connection between his plants and a sense of position within the broader urban area. He is a self-employed forty-two-year-old man, and he discusses the role of plants and sharing perennials within the context of first-time homeownership and neighbourhood:

It was my first house and my first yard – I didn't really know what to do. You get kind of overwhelmed, you don't know what to do first. So I just let all the weeds grow up and then by the second summer I was here, I decided I want to change this out and do some work … I learned a ton from my friend Susan, and she taught some other dudes who live around here, too. She sort of paid it forward, so then I thought well I should pay it forward, so – that was where the plant exchange and working with some of the ladies of the neighborhood came from.

For Jack, these "neighbor ladies" welcomed him into the fold of neighbourhood gardening and now also provide a means, by sharing his plants, through which he can translate these skills and experiences to "other dudes" who might be overwhelmed by their homeownership experience. He sees himself connecting with other men in the area as he shares his own plants and expertise with them. He sees his position as a single male gardener as fairly unique, a position with the potential to inspire and educate other men like himself to take up gardening. Jack talks about sharing plants in terms of his own evolving expertise as a gardener enabled by the help of several "neighbor ladies" and their informal networks of plant sharing and exchange.

Experienced gardeners see sharing plants as a way to pass along unwanted plants to others, and they often don't consider plants from others very desirable unless they are unusual. As Jack showed me around his front yard, he pointed to a newly planted area with several unusual sedums and ground covers. He said, "This was literally all tiger lilies when I got here, and I was just on a mission to get rid of them. I put some out on the boulevard with a FREE sign and gave the others away to some friends a few blocks away." For Jack, common tiger lilies and hostas conjure up images of yards counter to a more progressive, newer style of gardening to which he aspires:

I totally appreciate them for what they are, they're great for certain shady and sunny spots. But I usually find plants from the neighbor ladies. We do a casual plant trade, and I say yes cause they're almost always pretty unique plants. And then in turn, if they want anything I have, I'll split that up, or if they want some vegetables from my garden, I'll share them.

Sharing plants entails anxieties and uncertainties about the trustworthiness and degree of interest of a given plant, especially for more seasoned gardeners. One of the labours of commoning through plant exchanges and sharing, especially on the part of skilled and experienced gardeners, is a quelling of this anxiety through loose but sustained ties between people, yards, and plants. Barb remembers her experience forty years ago when she talks about her broader aversion to buying plants for her yard:

Almost everything you see here is stuff we've traded, so I've not bought things … We didn't have any money when we first moved in! We were busy making the inside of the house livable, and with the kids and all the expenses of a young family. I tell people, "You don't have to make a big investment, because you can trade with people and get things that work in their yard ..." So everything in this yard came from someplace else, came from somebody else, which, as I said, is kind of my philosophy.

Barb folds geographies at several scales within this commitment to plant exchange and sharing rather than cash investment: the micro-geographies of adjacent and nearby yards, and "inner-city" skills and qualities as embodied in the hardiness of certain plants. Barb is an accomplished gardener, and currently has been helping with two relatives' yards in other parts of the metro area, each of which she considers to be relative "blank slates." She told me what she has told them as they get started trying to decide on plant varieties and what they like: "We can get things that grow in neighbors' yards. If it works in their yard and it's two blocks away, it's gonna work in your yard!" She laughs and says, "The plants are gonna have to survive – and frankly, that's my philosophy, too – I say, we're an inner-city garden with inner-city values, so if you're gonna make it, you have to really get out there and scrap and duke it out, and say, 'Okay, I'm here!' So the garden has to be that way, too." Barb's understandings of a hands-off gardening style resonate with her understanding of her place in the broader urban landscape. Sharing plants and knowledge is one powerful form of exchange, supporting a thriving sense of life in common within particular biophysical and social geographies of everyday life.

Knowing Nearby Yards

Along Barb's block in North Minneapolis, neighbours share a deep and situated knowledge of one another's yards and plants, shaped over decades. This makes for an exception to some of the ambivalence or uncertainty about shared plants. Barb told me,

Well, we've almost all been here for 40 years, we've lived here for a long time, so we kind of know what everybody else's yard is like, who has what and what we can share. Like when Kay was moving a lot of her stuff to the community garden at the end of the block, we'd all say, "Okay, when you get to that one, I want that one!" So people would kind of know.

About thirty years ago, vocal residents approached the city, concerned about cars speeding through stop signs. They suggested that a small park could be built on the site of a large open intersection, closing off the road in the hopes that the park would act as a traffic diverter. A dedicated handful of some of these same people still living on the block have been tending and shaping the space over time and now cultivate several small gardens there. This area in North Minneapolis has relatively large homes and lots, first developed in the 1920s, and subsequently inhabited by waves of upwardly mobile Jewish residents, then primarily middle-class African Americans from the 1960s to the present. The park illustrates some of the ways micro-geographic proximities and adjacencies come to matter in an urban commons through material and practised linkages with privately owned spaces, as well as the multiple ways those involved experience these commoning practices. Involvement in the garden varies – as a way to make something for others in the area, as a means to fortify and reform boundaries and buffers, and as a space for creative expressions and enjoyment of the embodied practices of gardening.

This shared public space highlights the ways commoning practices weave together rhythm and time with private yards. There are no formally assigned plots, in contrast to many other community gardens – the vegetable area is a free for all, with whoever is interested planting what they like.

Waves of interest and involvement in this garden come and go, from year to year and season to season. Each year, the block club has a meeting so those who are interested can get together to talk about gardening plans for the park. The planted areas remain quite informal, and change from year to year. Barb told me her philosophy about these community projects is that "the right people always show up," and she explained that some people's interest in the community garden and their ability to be involved has waxed and waned over the years. As those most engaged in the garden get older, these shifting capacities for doing this sort of "community work" may become more and more significant in the maintenance and continual reinvention of the park.

The residents most involved are veterans of a variety of community projects, initiatives, and programs – and all share a sense drawn from these experiences that the less something depends on one or two key people, the more sustainable it will be. Although, at the moment, they each are involved closely with the garden and contribute a lot, the ideal in the way they

talked about such communal efforts is to have a diffused sense of obligation and responsibility. This is a concern, as the primary people involved in the community garden are over sixty and semi-retired. The physical capacities to keep up with the gardening tasks will increasingly inform how and how often these inhabitants are able to cultivate the community garden, in addition to their own yards. At the same time, their years of experience enable a kind of ease and skill that shapes and responds to the plants and space. Commoning entails embodied labours and abilities of commoners to engage with these changes as communities age.

The commons of this park/garden is linked materially and socially with nearby residents' yards, through the practices of neighbours moving plants. The park/garden has become a repository for unwanted plants – a destination to share plants that people may be phasing out of their own yards and gardens. Here, the excesses of cultivated private yards become the plants that shape the commons. Kay has made major additions of perennial

plants to the park in the process of reworking her front yard gardens the past two seasons. This transplantation and the subsequent necessary cultivation – weeding and watering – has become embedded in Kay's routines. She wanders down to the garden early in the mornings to weed. "There are some people who can't stand weeding, and this – I can spend hours here, weeding." Kay bent over to pull at creeping Charlie plants. She continued weeding: "Sometimes we end up meeting here. And someone will be weeding, and someone else will come down, and we have a little coffee klatch." Her hands grab at the low plants. "Or the other morning, I was here at seven, and our neighbor was walking to church – so we had a morning conversation."

The sense of time people experience when in the garden is part of a rhythm – a daily routine, but also a temporality outside of regulated time. What I later learned is that another resident, Ann, has done more with the community garden in recent years as a way to cope with her husband's faltering health – it has become a destination close by but also a place she can lose herself in the physical motions of being outside, digging, weeding, and watering, as well as making social connections. This garden serves a similar function for Barb and the others. Barb regularly waters the gardens, unfurling a long hose early most mornings from her yard along the sidewalk to the community garden. Kay told me, "A lot of us, we wander over in the mornings, and start weeding or doing whatever, and suddenly it's two in the afternoon and people are wondering where we are!" This common rhythm of the day resonates with longer life rhythms of retirement and time away from paid work within the lives of these neighbours. Further, there is the shared sense of time outside of time. All of this resonates further still with the plants' own rhythms of growth and change over the seasons. These nested resonances enable the cultivation of this common space and its shared sense of meaning to those involved.

Experiences and labours of commons vary, even among the commoners most involved in commons' production and cultivation. What labours may be more communitarian in nature for Barb and Kay, are, for another neighbour, a way to tame and keep at bay the wild edges of gardens interfering with her own yard. Eve, a newer resident of about five years, has transformed her front yard into a formal garden, and she devotes hours to creating and maintaining elaborate blooming perennials, topiary hedges, and ornamental trees. She became involved in working on the community garden primarily because a shaggy wildflower garden within it had infiltrated her yard with seedlings, wreaking havoc on her meticulous gardens. Eve's experience is a mix of frustration with others for not following through on their visions and a resistance to loose or unplanned gardens that complicate the ability of city workers to mow the grass. At the same time, she also clearly enjoys the aesthetic challenges of working up a "new" space. She was careful with her words when she told me about how she first got involved and her activity reworking the garden area closest to her house,

It was all outta control and all these seeds are blowing all over the garden. And it's a problem because, you know, people want things, but they don't really want to do the work. Or they're not able to. So – because this is so close to my house, how can I, um, harmonize with what's going on here. And have it be less maintenance, too.

She went on to describe the plants she is putting in – some purchased with neighbourhood funds, some transplanted from her yard as a means for aesthetic continuity. Maintenance and legibility are major concerns for Eve and her take on the community gardens:

> I told my neighbors, I'm sorry, but that community garden is gonna have to take care of itself, I have enough going on. And this new area is gonna be really beautiful, and it's gonna take not that much maintenance, and then the rest of the community is gonna have to figure out what they're gonna do with these spaces. Now, Kay has done a lovely little thing here [pointing], but then she has to come in and weed it, and do all that. So I don't really know exactly how long that will happen. I think the design of the garden has been really disturbed by all this busyness … There's not a clear delineation around some of the planted areas for someone from the city who's coming in with a riding mower and mowing the grass in ten minutes.

In this commons, cultivation practices are varied, multiple, and shifting, and are centrally tied to the broader contexts of commoners' lives. Eve's comments also rest on the position of the park within the broader institutional context. Its position within larger formal institutions is important for the way people feel about their own contributions of time and effort. It doesn't define how they see the potential of the park, but the recognition does strengthen the sense that it is not an individually driven project, and that it might survive over time. Each year, the block club has received a small grant (about $500),[12] and Barb told me, "We get little bits of money that comes along. So you don't have to spend your own money; you feel like you're part of a system and the system is supporting it." Although the amount provided by the "system" is fairly minimal, it is a meaningful recognition to those directly involved.

Plants are the primary means through which the practices of cultivating commons take shape and come together – through plant bodies as they move and are moved from place to place, as well as how their material characteristics are understood and made meaningful. All of this sharing and trading takes place within a broader field of high plant mobilities within and between yards. This mobility might be somewhat surprising to non-gardeners, as plants are so often seen as stationary in space, but, again and again, participants who were active gardeners told me about moving plants, often in great detail – around their own yards, and through the origin stories of particular plants from friends or neighbours, or found for free. In addition, the garden/park calls attention to the minor but important role of state recognition, and the legibility of communally produced green space. For passers-by, the small park is a place through which to walk on the way to other places, maybe to stop and sit, to chat. All of these neighbourly activities take place there on a regular basis. As the older generation instrumental in the creation and cultivation of the park/garden moves on, newer residents may understand it less as a commons produced and maintained through collective action, and more as a public park. This may be the end of the kinds of commoning practices currently at work in this space, or perhaps the potential for new forms to take shape.

Conclusion

Certainly people may be interested in communal cultivation practices precisely because the constant maintenance and incremental improvements stabilize and increase property values across a neighbourhood. But is this the whole story? What are the motivations and impacts of these kinds of practices? How do people explain them? What do they mean, and for whom? In some ways, the question of the extent to which these ordinary landscapes are understood to be sites of crushing capitalist exchange value or potentially liberatory sites of experimentation and common good misses the point – and is, in any case, impossible to determine. Seeing the diversity of yards and yard practices enables us to see logics of private property interwoven with logics of commons and commoning. Yards may not be singularly one or the other of these logics – instead, these logics enable one another in lived experiences. I think about these questions and experiences in the spirit of Gibson, Rose, and Fincher in their 2015 *Manifesto for Living in the Anthropocene*, in which stories have the potential to "enact connectivity, entangling us in the lives of others"; through making sense of the world in the stories we tell, the book argues, we are making possible worlds.[13] By looking for the common life within the lives of urban yards and finding a diverse range of spaces and practices, we can add to emerging conversations about the nature of urban commons and commoning. More broadly, we can think with greater richness about urban geographies by making legible some of the diverse practices and experiences in these often overlooked yard spaces, as well as the singular ways plants shape and are shaped by social relations at several scales.

Private property in these cases sets up possibilities for common life, but also constrains it. People have the potential to form meaningful connections across property boundaries in and through yards, but access to these urban landscapes is primarily limited by homeownership. Yards do provide the potential to reimagine urban landscapes, neighbourhoods, and everyday encounters. But, as with any commons, this is a textured terrain of openings and closures. Micro-geographies of adjacency and proximity between common cultivated spaces and individual yards reveal an interconnectedness that crosses legal categories – through plant tissues and bodies, and also through everyday commoning practices and rhythms. In the above cases, the ways people live with their own and others' yards over time involve a variety of practices and understandings of property. Looking for commoning in and across existing yard spaces and practices might enable seeing concrete foundations for the kinds of socio-political transformations implicit in much of the literature about urban commons. This type of seeing requires thinking about yards beyond singular lots, extending analysis along lines of flight toward how property might be otherwise.

The common life of yards entails specific geographies shaped by the lives of plants, rhythmic and routine encounters, as well as a situatedness within broader urban contexts. Parallel with the ubiquity of yards is the ubiquity of how we tend to see territory and property relations.[14] In both cases, the familiarity surrounding us obscures the dynamics actually happening here – and the recognition that there is so much potential for change right under our noses. The circuits of plant tissue described above, and the ways residents experience these, form intersections of ownership, care, cultivation, and land. In these intersections, commons are cultivated beyond shared territories or public spaces. Commons can be mobile and carry the shared life force of plants themselves. In yards, commons grow.

Yard Futures

To use the world well, to be able to stop wasting it and our time in it, we need to relearn our being in it.

URSULA K. LE GUIN, "Deep in Admiration"

Although the spaces and routines of daily life may feel sedimented and static, they are anything but. In large part, ongoing processes of making give shape to these spaces and routines – the production of space and society. The way we tend to think about "the everyday" flattens together many days. It comes to feel like a stable, bounded category of the familiar. The spaces and routines we inhabit give shape to this accumulation of the everyday. These spaces and routines produce and maintain senses of self and other, geographies of here and there, rhythms of past and present. If we understand more about how the meanings, practices, and spaces of everyday life come to feel so settled, perhaps we might find further ways to unsettle and remake the world. This possibility for change is at the heart of the perpetual hope of many theorists of the familiar. The focus of this book reflects the interplay between habitats and inhabitation, revealing potential within everyday encounters. The yards we have encountered tell a story of this similarity and difference.

Rather than adopting a universalizing narrative of human experience, the preceding chapters have revealed rich variation in the everyday lives of yards, which we rarely see or appreciate – variation in forms, uses, and meanings. Through the lens of sustainability and changing environmental management, we have seen planners and environmental advocates increasingly mobilize habitat changes for better water quality through stormwater management, expanding wildlife habitat, and increasing local food production. Sustainability plans and goals have been imagined through quantifiable measurement, as well as universally transferrable material best practices. These plans, goals, and practices have relied on yards as bounded and owned territories, underscoring individualized neoliberal subjectivities of owner, customer, and manager. In Minneapolis, yards have been one site where such plans and practices meet the nitty gritty of hands digging in the earth or water draining from a downspout. Rain gardens are made and maintained, with the help of neighbours. Trees grow again where they were lost. These examples contribute to stated sustainability goals, but is this enough? These physical features of yard habitats also include how people live with them.

Inhabitation – how people live, and live each day – supplements and complicates these measurements of more sustainable urban habitats. The world-making in which people are engaged every day, however repetitive, is a repetition with the possibility of difference. Simply the variation in yards we have come to know in this book shows us that there is more to the familiar than is usually considered. People have changed and made their yard worlds over long periods of engagement through shaping, rebuilding, cutting back, seeding. Over time, particular yards are produced, as are dispositions oriented toward those yards. But these ways of being with yards are not solely human. In addition to the significance of human practice in the everyday, it is the inclusion and consideration of encounters with others – human and more-than-human – that transform built environments into phenomenal ecologies. The view of yards through this lens centres how people live with architecture, urban design, and urban geographies. Here, the flesh of the body lives with the flesh of the world.

And yet, even these views of yards benefit from additional angles. The focus on these immediate embodied encounters can't quite explain the atmospheres of relation between self and other, home and yard, yard and city block. We have seen the deep embodied attachments and meanings of yards, in how people tell stories to understand their pasts, their families and children, their neighbours, their pets, and their own geographies within the broader urban environment. Yards are one way in which the people in this book understand themselves. This is the yard as a felt realm, of sensing between bodies human and more-than-human. Experiments in knowing these relations have unfolded in this book through the telling and retelling of yard stories told in place. The practice of looking for and capturing yard affects became experiments in pairing images and text as a photo essay in the preceding chapter. The lens of affect, emotion, and care showed us how yards are places of attunement to affective atmospheres.

These atmospheres circulate between notions of self, other, and nature that are shaped by urban history and geographies of capital. The familiar often falls back to a way of living with yards that reinforces abstract notions of property and ownership. More often than not, inhabitants live within these bounds and practise their yard in order to maintain them. But we have learned about variations in the socio-natural relations of property – how people restructure their relations to neighbours through yard spaces, and how plants and animals disrupt and transgress bounds of ownership. Openings in fences are made, territories shared. Plants jump the fence. People divide plant tissues and move them all across the city. Gardening expertise circulates across city blocks, most often led by older women who know a thing or two about nurturing life in its many forms. These are on-going relations that shape and reshape yard spaces and the neighbourhoods that are a connection between parcels of land and people. This is practising property over time.

Taken together, this book shows how habitats and inhabitation of yards must be understood together. The yard is a convergence of structure and practice. It is bounded and yet its inhabitants contain the potential to leap across those same bounds. There is more variation and potential in yards than is easily seen, or is readily acknowledged and mobilized by urban environmental advocacy and management. Yards are deeply felt sites of kinship with others – human and more-than-human. The ways people come to know their yards and tell stories about them contain atmospheres of attunement and connection to others. Sometimes these affects shape how yards as property are practised in new and unexpected directions. Together, these views into yards and their inhabitants reveal aggregations of connection, attunement, proximity, and care. The following sections discuss how a fuller account of living with yards includes implications for urban inhabitants and also experts, managers, and advocates. Given the challenges of the contemporary moment, these insights might better inform how we live with yards.

Potential within Everyday Encounters

The way people talk about and experience yards is centred on care and cultivation – self-conscious environmental care labours, but, perhaps more significantly, desires to care for, and simply connect with, others. This happens through varying capacities and skills, shaped by life histories and past experiences in place. It is animated by social relations, especially the ways in which people circulate knowledge and materials at the scale of the city block as well as through social networks with broader geographies across the city. People care deeply about, and for, their yards. We have learned how people constantly devote time and energy to them. The yard is one place where many people experiment, get outside the confines of their houses, enact creative visions, and make and sustain connections to others.

Plant a plant. Select a spot, dig in the dirt, hold up the stem with one hand while filling in the gaps. Press the earth, it presses back, water ring around stem. Let your fingernails get dirty because you can't stand the feeling of gloves. Because you want to touch the dirt itself. You want to touch your yard. Get a blister from your efforts. Feel frustrated it doesn't look the way it should, plants languishing. Feel failure. Where to even begin? Be uncertain about what to plant, and where. Struggle with hose and other tools, if you have them.

Look out this window. See what is there and what might be, also what was. Wonder about what it needs. Notice big and small things. Attune. Tune out everything else for a time. Watch and wait. Watch and plan. Watch things grow.

For urban and environmental planning, all of this means small changes may make big differences. Changes and their impacts will vary within a city, depending on people, neighbourhood relations, and decision making at local scales (for example, real estate development pressures). Identifying and being responsive to the social dynamics at play in a given area of a city, for example, would go a long way toward unsettling the universalizing property-owner model for environmental interventions and adaptations. Similarly, pushing toward more reparative and just planning processes and outcomes must include accounting for historical disinvestment and displacement. One of the best ways to do this is in working with community. The upsides of this approach can be seen in the shift of the rain-garden organization in Minneapolis from blanket installation of large numbers of rain gardens to facilitating more localized and supportive programs alongside neighbourhood groups already at work in the area.

For urban inhabitants, everyday environmental care is important and significant – not just in ways that can be measured, but socially and politically. Yards make possible particular kinds of social relations – passing by, stopping to talk, experiencing others' aesthetic and skilful cultivation, noticing changes in capacities for care and maintenance. This kind of barometer is not just visual but becomes activated through inhabitation. Discovering them in their nuance requires the kind of detailed territorial approach that enabled this study. For future directions in yard study, collaborative partnerships with community organizations and activists would provide important additional perspectives on these social and political dimensions of yards, such as immigrant experiences with yards and home ownership, urban and suburban community land-trust initiatives, or informal economies circulating through yards. Do urban inhabitants, planners, engineers, or gardeners recognize the potential within these everyday encounters? Perhaps through better seeing these encounters, and the variations in them, already right under our noses, the inherently caring labour of living with yards might be directed toward more egalitarian and just ends. Although many live within systems and landscapes organized around territorialized private property bounds, there is no reason these bounds can't be undone in small and big ways. They are already being undone, as the many stories and experiences of this book have shown. The world-making within the yards we've come to know in this book shows how much more care and cultivation goes into living with yards. Other yards are possible, just as other worlds are possible.

Mow the grass. It's there; it needs to be maintained. Or don't; it doesn't. Dig it up in places. It shows you and others about your self.

Pull the weeds from their footings as you talk with someone on the phone or to a dog. Feel the resistance and release. This is best done fairly soon after a rain. Cast aside the small weed bodies. Their white roots will be thin and bent amongst the green.

Water the plants, the trees really. Smell the water soak into the earth. Drag the hose behind you as you go, feel its weight. Make a spray with your thumb, which will get cold. Or set up the sprinkler on particular days. Or use a watering can, carry it heavy then light. Repeat. Hope for rain. Water from a hose is never as good.

Walk around the house at the end of the day. Notice things that need to be fixed, changed. Notice what plants are doing differently from the day before, week before, year before. Admire them as you scrutinize. Are they happy?

Storytelling Is Critically Important

When we try to explain the world, we make the world.[1] I have approached yards through visits in place and the stories that unfold there. Touching plants, walking through the grass, listening to yard sounds, smelling blooms. How we know these familiar spaces such as yards needs to take all of this into account, including registers that exceed familiar modes of expert and scholarly representation.[2] So much of what is meaningful about yards to people has to do with attachments and engagements with more-than-human worlds, in temporalities outside of time. These attachments are felt and practised. Yards make us, just as much as we make them. Yards allow us to glimpse the kinship between the sensing body and sensed things that makes (an)other kind of communication possible. As the stories in this book reveal, storytelling as a mode of research and relation can begin to capture the power of this kinship.

For planners, this understanding means making space for stories to affect the future. This is not about listening as cursory performance, where constituents tell stories and planners nod along while plans proceed unchanged. Planning and imagining urban environmental futures must take seriously lived experiences as understood and told through stories. For inhabitants, the many experiences in this book might be seen as an invitation to consider the stories that people across intersecting axes of social difference all want and need to tell one another about our lives. Yard visits were about yards, yes. But what matters is the way these spaces become meaningful through the doing of maintenance, the imagining of redesign, or shifting things around based on growing expertise and experience. Through stories, we learn about one another. Skills and aspirations. Frustrations and histories. We learn about homes and timelines of change through stories of renovations and repair. Not enough money or time or experience. And stories are not equal, in their reception or in their impact. Stories of spaces like yards intersect with social differences and geographies. Those in powerful positions tend to envision yards through a normative lens, bound by tidy aesthetics and the straight grid of property lines. For decision makers in Minneapolis, stories from South Minneapolis and North Minneapolis are heard differently. The stories that unfolded from yard visits were shaped by all of these social and geographical differences.

Planners might try to understand such practices as a resource, not a barrier. To do so means being open to diverse kinds of gardens, species, and spaces. And the diverse socialities of yards, garages, and alleys. For inhabitants, this openness invites reflection about understanding their past in relation to the present. What has informed gardening skills and failures? What sorts of possibilities emerge, for example, from intersections of the privileges of whiteness, growing up in a milieu of middle-class segregated stability, and familiarity with this climate? These experiences simmer under the surface for white participants – informing what is taken for granted, what feels natural about yards. Many assumptions are embedded in how yards are experienced and practised – varying levels of familiarity with owning a house, job security, being able to make plans for the future, upbringings with more or less precarity, in which skills such as gardening made life more abundant. For white upper-middle-class participants in South Minneapolis, privileges and positions along social and economic lines were rarely uttered out loud, if even acknowledged. For more working-class participants in Northeast Minneapolis and North Minneapolis, people often described their neighbourhoods relative to South Minneapolis, or in relation to "the suburbs" – shorthand for more affluent and anonymous areas. Hardships of mortgage payments and the distance between aspirations and costs of projects constrained possibilities more in these neighbourhoods. But they also gave rise to responses based on skills and the socialities of shared expertise. In North Minneapolis, African-American inhabitants emphasized a deep sense of home, rooted in social connections with neighbours extending across generations. Long-time white participants on these blocks emphasized their commitments to staying in the area through thick and thin and over decades, with the assumptions they could move elsewhere in the city, but instead chose to stay. In both cases, as we walked through yards, neighbours often called out greetings, grown-up children of friends stopped by, young kids were known and welcomed, families on the block were described from grandparents to grandchildren and great grandchildren.

Remember past yards. Stake the raspberries like your grandfather did. Move your great grandmother's Siberian iris from house to house, yard to yard, decade to decade. And how did your mother use the space in front of that Chicago home? What kind of roses did she grow?

Whose yard is this anyway? What does it mean to own something? What does this take? And from whom? Find the answer right under your feet, the land.

Think about how your city yard now raises a city person, not like the farm person your parents' farm garden grew.

Changing Habitats, Changing Inhabitation

Yards often entail long-term engagements over decades, because home ownership is shaped by ten-, twenty-, thirty-year mortgages and related financial arrangements. People grow with their yards, and vice versa. These long engagements are marked by rhythms – by growth and decline of more-than-human organisms such as trees, by life histories, and by changes in weather and season and those resulting from renovations and repairs. There is also an accumulation to yard rhythms, through iterative actions and practices. Sometimes those actions and practices are a way to step outside of, or disrupt, the passage of linear time. For example, people lose their sense of time by weeding. Tasks associated with seasonal changes punctuate linear time in that they must be done within a short window. Right now. Seasons return again and again. Until perhaps they do not. Our era of climate change continually invites new attunements to the ways our living surroundings respond to seasonality and its changes. Just as with knowledge that builds up in place, from year to year, rhythms of expectation and change become more and more familiar the longer people live with their surroundings.

Yards are about much more than gardens and gardening. But cultivation through gardening is at the core of many yard practices. Gardening is a practice without a lot of limits – not dependent on formal education level, age, income, or wealth. There seems to always be a way to make a garden work, even in the most unlikely places and times. Accumulated knowledges and experiences of individuals and communities with shared histories make possible the innovations in these everyday practices. Even as gardening becomes caught up in trendy visions of the future prone to green washing, the accessibility of gardening enables everyday creativity to take root and flourish, which in turn continues to hold potential for giving shape to social change. People try it out, cobble it together, take it apart and rebuild it. I saw this in the yards of people whose mortgages were upside down, who traded plants in alleys because they couldn't easily afford new hostas, in inhabitants who found materials and built new spaces always with an economy of necessity in mind, using what was readily at hand.

For environmental experts, it is crucial to take into account how people live with habitat changes and prescribed best management practices like permeable paving, rain gardens, and composting. From talking with experts and inhabitants about yards, it is not immediately clear how people actually live with such changes in habitat and related maintenance. The long-term success and effectiveness of many physical adaptations like green infrastructure features is not yet known and may depend more than anticipated on the many mundane and important relations of inhabitation found in stories such as those in this book. Planners might consider how changes in habitat relate to experiences of time and rhythm. Cities like Minneapolis continue to change in density and affordability, and urban forms are changing. We might collectively set priorities to consider the kinds of habitats that could foster the deeply meaningful rhythmic inhabitation of living with others, human and more-than-human. Can designers imagine these beyond the spatial configurations of a yard surrounding a single-family home? Or beyond the shared green spaces of parks? In which ways might balconies, patios, or garden plots, for instance, also provide moments of response and connection to rhythmic inhabitation?

Ultimately, the question of changing forms of the city through adaptation measures and redevelopment is often organized around contested densities. Integral to questions of density for urban forms like Minneapolis are connective tissues such as yards. Culminating in 2018, the city council developed a relatively progressive planning vision in the Minneapolis 2040 Plan.[3] Most notably for housing and neighbourhoods – and after several years of contentious debate, community engagement, and extensive revisions – the plan eliminates zoning for single-family residential land use and instead envisions neighbourhoods with increasing density and innovative built forms. Policymakers articulate the broader goals of the 2040 Plan to centre around racial equity and redressing generational wealth and housing gaps across Minneapolis geographies. In terms of built forms and neighbourhood design, these policy changes amplify trends already underway to incorporate more accessory dwelling units and multi-family housing into residential areas, to reimagine green spaces in smaller and more interior forms, and to allow and encourage more mixed land uses. The specific impacts for yard spaces across different neighbourhoods remain to be seen. But for a city with so much connective green space and urban tree cover as Minneapolis, the plan points toward more density and infill development on small and medium parcels of land. Naysayers may emphasize encroachment on important open green space and losing the character of streetscapes and familiar neighbourhoods, but the stories within this book have the potential to push collective conversations about green spaces such as yards toward the socio-natural and socio-spatial relations embodied within them.

In other words, what kinds of habitats and practices of inhabitation might help people further develop the best of yards, and minimize or even redress the worst? This may be a critical moment to open up access to the transformative relations of yards more equitably and to more inhabitants of the city. To answer these questions will require designers, planners, and environmental managers to think and work in new ways. Being guided by the emerging principles of design justice, including the wisdom of those most affected by ongoing systemic environmental injustice, remains the minimum starting point. Regardless of these interventions and plans, inhabitants will make these new spaces meaningful through everyday practice and experience. Think of what might be possible if these threads could be woven together.

Talk with a neighbour across the yard. Call out. Reach out. Wave. Throw something back to them. Offer them something grown. Make friendly, neighbourly, pleasant conversation. Affirm what they say, what they do. Most of all, admire something alive together.

Move through the edges of others' yards. Make openings in fences, widen the cracks. Take a visitor into another backyard that you've known for forty years. Fling open your gates to those who want to dig in your dirt. Share your sunshine. Provide water to plants others planted. Divide your hostas and give them away, circulate their tissues in ever widening circles. Refuse to hoard this little paradise.

Layered on top of these recent changes to city plans are the recent realities of the novel coronavirus COVID-19 pandemic, which emerged in North America in 2020. Especially for those willing and able to isolate at home, and otherwise limit more public activities, the significance of access to outdoor green spaces could not be more clear. The use of city, regional, and state parks skyrocketed during the spring, summer, and fall months of 2020. Outdoor green spaces have provided Minneapolis inhabitants with room to move, places to socialize at a distance, and simply a destination beyond the confines of interior domestic spaces. For those with ready access to yards and resources like time and money, the pandemic time has intensified uses of yards as social spaces. Front yards increasingly feature small paved patios or decks where people can gather safely outdoors. People have furnished yards with fire rings and gas heaters, outdoor seating, and various tent structures for protection from rain and snow. The pandemic and the collective response to it have hollowed out many of the most meaningful shared and public places and activities of city life, leaving some to question whether higher urban density is worth the trade-off of more limited private outdoor space. In response, some have made major life changes by relocating to areas with relatively more open green space. Cities such as Minneapolis may seem increasingly attractive to residents of more densely packed coastal cities. Rural areas attract city dwellers. However, racialized housing segregation and gaps in wealth remain high, and affordability a major obstacle to fairer housing in Minneapolis – as the 2040 Plan and other long-standing organizing efforts show.

The pandemic time has become only the latest crucible of deeply interlocking systemic crises: racial injustice, climate emergencies, major extinction events, conflict and displacement, widening disparities in income and wealth, and the erosion of functioning governance. The Black Lives Matter movement has led diverse collective discourses and kept sustained attention on the ongoing systemic racism in every facet of American life. With solidarities emerging in new ways, movements such as Indigenous-led organizing for water protection, in the face of fossil fuel extraction, and the Land Back campaign question the foundation of property ownership that underpins all dominant relations to land. As these movements so clearly show, urban habitats must be seen as deeply flawed manifestations of ongoing global geographies of white supremacist dispossession and exploitation. Collectively, those involved in shaping urban environments – whether through everyday practices or formal professional roles – must attend to the long reach of white supremacy and racism at the foundation of geographies of home ownership and access to yards, and in how different communities within a city or metropolitan region may be understood in relation to environmental planning and management.

Sit. Notice the world around you. Watch it. Swat at insects. Drink something cold or hot. Keep watching that world. Notice the squirrel that watches you. You have chosen a spot. On steps. On a chair. Next to a tree. Under an overhang. Do this at the end of one day or the start of another, or both. Feel air shift across your skin.

Smell the blooms. From inside your kitchen window. Through the screens, the scent reaches you and tells you things. Summer is here. A time. A place. This lilac.

Living with Care and Cultivating Life

How can small parcels of land, including yards, even begin to be a part of the reparative work that is so needed to redress all of these violations? Some of the answers, subtle and almost invisible, pulse in the yard stories of this book. Circulations of plants, sharing expertise, reciprocal care attuned to the nurturance of life across differences. Yards are sites of forging intense emotional, embodied, and social relations, shaped by urban policy and history – but not limited by them. In everyday encounters, people experience these small patches of land not as abstracted property parcels, but as earth and sky to be shared with others, as home.

Stretch the boundaries of yards to breaking. Imagine future yards. What more can a yard be? What else might a yard make possible?

As the world that we know attests, however, these answers are not enough. There are many more experiments to be tried. There is so much more radical potential in these sites of reciprocity, repair, and remaking the relations of the world with each new inhabitant. Who gets to, who wants to, and how must we practise yards otherwise? We can make opportunities to engage more insistently with the desires of care and cultivation with others that permeate this book. We can make opportunities to experience rhythms of inhabitation and the affective ecologies of yard terrains in spatial arrangements beyond conventional private yard territories. We can make understandings and practices of property beyond survey lines, instead reaching down into the earth and up into the sky. The question remains about how we will live with these habitats, even deeper in admiration, as Le Guin writes. But for a host of reasons, it is essential to try.

ACKNOWLEDGMENTS

HEARTFELT GRATITUDE to all participants of this project, for so generously welcoming me into their yards. At outdoor tables, under birdsong, and surrounded by living things, you surprised, delighted, and challenged me. It has been a great honour to learn from your expertise. Thank you. Thank you also to the designers, advocates, and policymakers whose work shapes yards in lots of ways, and from whom I learned a great deal about making, maintaining, and imagining urban habitats.

Support from the following institutions enabled me to research and write portions of this book, for which I am grateful: at the University of Minnesota, the Department of Geography, Environment and Society, and the Institute for Advanced Study; the Urban Studies Foundation and the University of Glasgow School of Geographical and Earth Sciences; and at the Rhode Island School of Design, the Department of History, Philosophy and the Social Sciences. To all the students with whom I've had the privilege to learn, thank you.

This book benefited from expert help at crucial junctures. Many thanks to Kathleen Kearns, whose developmental editing was invaluable in advancing the book. Thank you to everyone at McGill-Queen's University Press, especially editors Jacqueline Mason and Kathleen Fraser. I am grateful for the expert copy editing of Barbara Tessman. Thank you to the reviewers whose generous engagement with the text improved the book significantly.

This book has travelled with me across disciplines, degrees, institutions, professional roles, and continents, even when I did not realize it. I am grateful to early mentors who each shaped the research and writing foundations of this project for the better, even when none of us knew it. Thank you, Bruce Grant, Rachel Merz, Alison K. Brody, Jeanne Arnold, Margaretta Lovell, and Kathleen Moran. Special thanks to Paul Groth, who guided my first research into backyards when I was a wayward architecture student, and who made it possible for me to see the built environment as it is lived all around us. As mentors during my doctoral work and beyond, Katherine

Solomonson and John Archer, saw the connections across disciplines I've tried to make in this book well before I did, and I am grateful for their encouragement and advice all along the way.

Many colleagues became friends and mentors as I found my way into geography. Thank you to George Henderson, Eric Sheppard, Rod Squires, the late Roger Miller, Karen Till, Teresa Gowan, Valentine Cadieux, and Tracey Deutsch. Dona Schwartz generously offered her singular wisdom on ethnography and photography in early stages of research. Fellow travellers, I have loved learning from your insights – Heather O'Leary, I-Chun Catherine Chang, Omar Imseeh Tesdell, Chris Strunk, Renata Blumberg, Katie Pratt, Kristen Nichols-Basel, and Emily Bruce. Later stages of the work benefited from generative conversation and feedback from Hayden Lorimer, Damian White, Yuriko Saito, and Neera Singh, for which I am grateful. Special thanks for friends who have been near and far from this project, but always close to my heart, especially Alexis Burck, Reena Vaidya Krishna, Hannah Mody, Karen Mauney-Brodek, Laura Boutelle, and Heather McLean.

My deepest gratitude goes to Vinay Gidwani and Helga Leitner. I am indebted to their highest expectations and their tireless confidence in my capacities. I continue to be inspired by the examples of their own intellectual curiosity, integrity, and generosity to younger scholars. Their individual and collaborative wisdom brought out the very best of this book at each turn. That our friendships grew along the way is the part I cherish the most. Thank you.

Finally, I thank my family. I am grateful for the boundless love of my brilliant sister, Cornelia Lang, and her family. To David Peterson, my partner in everyday life. You have made this book, and much more, possible. To my sons, Anton and Leonard, you made some parts of this book harder, and some parts easier. You both continue to make all of it more meaningful. This book is dedicated to them, and with love to my parents, Gretchen and Jeffrey Lang. Every day with each of you is a joy.

FIGURES

People relaxing in their backyard; man in hammock, Hutchinson, 1899. Harrington and Merrill Families Photograph Collection. Courtesy of Minnesota Historical Society. / xv

"Life comes easy, June '64" (Mary Heaton seated on a chair in a backyard), 1964. Courtesy of Minnesota Historical Society. / xv

Storm water management at a Northeast Minneapolis residence. Photograph by Ursula Lang. / 28

Private notice about city regulations of dog feces, Minneapolis. Photograph by Ursula Lang. / 41

"I am a raingarden." Photograph by Ursula Lang. / 43

Front yards with rain garden in foreground. Photograph by Ursula Lang. / 44

Evening information session, Northeast Minneapolis. Photograph by Ursula Lang. / 47

"Where shall we start?" Photograph by Ursula Lang. / 79

Leaning toward plants. Photograph by Ursula Lang. / 80

Hand with flower. Photograph by Ursula Lang. / 82

"This is where I belong; this is what I take care of." Photograph by Ursula Lang. / 83

Front yards with sidewalk, Northeast Minneapolis. Photograph by Ursula Lang. / 85

"It's blooming to my satisfaction." Photograph by Ursula Lang. / 86

"I've eaten from this garden." Photograph by Ursula Lang. / 88

"Tomatillos just bust out of their husk." Photograph by Ursula Lang. / 89

Ground cover in the front yard. Photograph by Ursula Lang. / 90

Yard portrait. Photograph by Ursula Lang. / 91

"The way into my backyard oasis." Photograph by Ursula Lang. / 93

Bodies in the landscape, of the landscape. Photograph by Ursula Lang. / 94

NOTES

INTRODUCTION

1 Tessyot, *The American Lawn*; Jenkins, *The Lawn*.

2 But see the work of landscape historian J.B. Jackson, especially "The Popular Yard," and the essay by Paul Groth, "Lot, Yard, and Garden: American Distinctions." See also Grampp, *From Yard to Garden,* as well as Dianne Harris, *Little White Houses*. For analysis of suburban design issues related to open space, see Girling and Helphand, *Yard-Street-Park*. For an expansive analysis of the origins of the ideal American suburban dream house, see Archer, *Architecture and Suburbia*. See also Henderson, "What (Else) We Talk about."

3 See Jenkins, *The Lawn*, and Robbins, *Lawn People*. For a recent discussion of yard-related research in human geography and related fields, see also Lang, "Keep Off the Grass!"

4 Notable examples include Westmacott, *African-American Gardens and Yards in the Rural South* and Gundaker and McWillie, *No Space Hidden*. Both show, in different ways, how outdoor domestic environments continue to be central to the making of individual and community meaning through particular visual and spatial languages.

5 Art environments created in domestic outdoor spaces encompass works of construction, assemblage, bricolage, and decoration, and are sometimes considered folk art or outsider art. SPACES (Saving + Preserving Arts and Cultural Environments) Archives at Kohler Foundation in Wisconsin (http://spacesarchives.org/) catalogs, documents, and archives materials related to art environments all over the world, and includes art environments of the upper midwestern United States.

6 Within human geography, two notable works investigate such questions: Head and Muir, *Backyard*, and Robbins, *Lawn People*.

7 Studies of home, home place, and home geographies span wide-ranging approaches. For an introduction to geographies of home, broadly conceived, see Blunt and Dowling, *Home*. The inhabitation of these home geographies is shaped by relations between everyday life, architectures, and bodies – which are, in turn, shaped by social differences. For an emphasis on architecture, see, for instance, Harris and Berke, *Architecture of the Everyday*, and Rendell, Penner, and Borden, eds., *Gender Space Architecture*. See also Grosz, *Architecture from the Outside*, and Cheng, Davis, and Wilson, *Race and Modern Architecture*.

8 Analysis of high-profile examples such as the High Line in New York City high-lights debates about green gentrification; see, for instance, Wolch, Byrne, and New-ell, "Urban Green Space." For consideration of the ways in which greening is mobilized in municipal policies and projects, including urban gardening, in smaller cities, see Strunk and Lang, "Gardening as More Than Urban Agriculture."

9 RC100 Program, Rockefeller Foundation, "What Is Urban Resilience?," accessed 30 March 2021, https://resilientcitiesnetwork.org/network/. Similar meanings are embedded in the language of the UN-Habitat definition: "Resilience refers to the ability of any urban system to maintain continuity through all shocks and stresses while positively adapting and transforming towards sustainability. Therefore, a resilient city is one that assesses, plans and acts to prepare for and respond to all hazards, either sudden or slow-onset, expected or unexpected. By doing so, cities are better able to protect and enhance people's lives, secure development gains, foster and investible [sic] environment and drive positive change," https://unhabitat.org/resilience/.

10 Massey, "Vocabularies of the Economy."

11 MacKinnon and Derickson, "From Resilience to Resourcefulness."

12 With gratitude to Matt Hern for generatively analyzing this phrase in the context of the real estate industry in his book *What a City Is For*.

13 This sense of possibility lies at the foundation of Henri Lefebvre's major work, the three-volume *Critique of Everyday Life*. I draw on this insistence on possibility within Lefebvre's work, even while analysing the limiting structures and narrow concepts of everyday life. Related interpretations of everyday life emphasize the invisibility of gendered experiences, routines, and cultures of institutions, as well as traditions interpreting the built environment and architecture through cognate concepts such as *ordinary, familiar,* and *vernacular*.

14 See a fruitful discussion of this phrase, and more broad thinking and writing about what might be otherwise, in Pandian and McLean, *Crumpled Paper Boat*.

15 Gibson-Graham, Cameron, and Healy, *Take Back the Economy*.

16 Blomley, *Unsettling the City*.

17 Elwood, Lawson, and Sheppard, "Geographical Relational Poverty Studies."

18 With gratitude to Neera Singh for discussing with me this sense of belonging specific to yards.

19 See, for instance, the works of J.K. Gibson-Graham, *A Postcapitalist Politics*, as well as Roelvink, *Buildling Dignified Worlds*, and Federici, *Caliban and the Witch*.

20 Entry points to this expansive material and relational turn in geography and related fields include, among other works, Bennett, *Vibrant Matter*; Braun and Castree, eds., *Social Nature*; Gandy, *Concrete and Clay*; Haraway, *Staying with the Trouble*; Latour, *Reassembling the Social*; Whatmore, *Hybrid Geographies*; and White and Wilbert, *Technonatures*.

21 Stewart, "Atmospheric Attunements."

22 Merleau-Ponty, *Phenomenology of Perception*; Ingold, *The Perception of the Environment*. Also see Simonsen's recent renovation of geographic engagements with phenomenology, based on interpreting Merleau-Ponty's focus on fleshy encounters and embodiment, but addressing critiques of apolitical phenomenologies within geography, "In Quest of a New Humanism." Similarly, Ahmed argues in *Queer Phenomenology* that orientation and disorientation of and between bodies and surroundings make possible a politics that reorders relations. Phenomenology can and must engage with political questions of social differences, particularly if phenomenologies draw on and reflect on embodied experiences.

23 See Singh, "Payments for Ecosystem Services" and "Affective Ecologies and Conservation."

24 Garcia-Lopez, Lang, and Singh, "Commons, Commoning and Co-Becoming." Fundamentally, this is a pursuit of recognizing and expanding how we collectively conceive and practise being human, especially beyond narrow views of white colonial "Man." See also McKittrick, *Sylvia Wynter*, and Cadena, *Earth Beings*.

25 Lefebvre, *Rhythmanalysis*.

26 Here I reference an insight from Ingold, who upends the distinction between *artefacts as made* and *organisms as grown* by arguing that the processes of making and growing are actually not very different at all. Ingold writes of artefacts as grown, and organisms as made in *The Perception of the Environment*, 345.

27 I build on connections between the politics of storytelling, feminist geographies, diverse economy approaches, and environmentalisms, such as those drawn in Gibson, Rose, and Fincher, *Manifesto for Living in the Anthropocene*.

28 See Lorimer, "Cultural Geography," for discussion of interplay between representation, practice, feeling, and sensing, which has long been an area of interest for cultural geographers.

29 Head and Muir, *Backyard*; Arnold and Lang, "Changing American Home Life."

30 With many thanks to photographer Dona Schwartz for several in-depth conversations about photography and ethnography, early in the development of research for this book.

31 Martin and Pentel, "What the Neighbors Want."

32 History and digital mapping resources, many specific to Minneapolis but relevant across US cities and towns, may be found at *Mapping Prejudice*, University of Minnesota, at http://mappingprejudice.org. For further insights into the legacies of housing segregation in Minneapolis, see also the PBS documentary on the Twin Cities, *Jim Crow of the North* (2019), at https://www.tpt.org/minnesota-experience/video/jim-crow-of-the-north-stijws/.

33 This finding fits with broader regional cultural expectations about doing one's own yard maintenance and gardening, often emerging from dominant settler colonial perceptions of rural and farming "roots" in this part of the upper midwestern United States. Compare this with, for example, the prevalence of hiring

help for yard maintenance and gardening in a city such as Los Angeles, even for middle-class families. These differences point to the importance of cultural geographies and meanings of place, and especially to the uneven geographies of informal labour for those who can afford hired help for yards and gardening.

34 City of Minneapolis, "Sustainability: History," http://www2.minneapolismn.gov/sustainability/history/index.htm.

35 See specifically Ingold's articulation of post-Cartesian modes of living with more than human surroundings – what he calls a "dwelling perspective" directly in conversation with Martin Heidegger, Gregory Bateson, Maurice Merleau-Ponty, and others, in *The Perception of the Environment*.

36 The constant labours of keeping things at least minimally in some kind of order and shoring up physical features occupy inhabitants in world-making ways, though perhaps in less compelling ways than dramatic societal ruptures or even organized social movements. I first became interested in maintenance partly in response to what felt like a lack of theorization within geography about the way in which sociospatial relations pause, slow, and often repeat. I have drawn inspiration from the artist Mierle Laderman Ukeles, and her "Maintenance Art Manifesto!", in which she articulates two systems, Development and Maintenance, and aligns women and domestic work with the undervalued latter.

CHAPTER ONE

1 The meaning of *environment* is elusive, enabling reduction and simplification in particular directions. This "everywhere and nowhere" is also a critique often levelled at the ways in which *sustainability* has been taken up and mobilized. See, for example, Swyngedouw, "Impossible 'Sustainability' and the Postpolitical Condition." See also Krueger and Gibbs, *The Sustainable Development Paradox*.

2 Indeed, such ways of knowing have real consequences. For example, a recent study on the diverse nature of gentrification across Minneapolis and Saint Paul found differences in how the impacts of gentrification were understood and taken up in advocacy and policy making: Goetz et al., *The Diversity of Gentrification*. Quantitative findings were less likely to identify gentrification as having significant negative impacts, while qualitative research was more likely to point toward hardships intensified by processes of gentrification.

3 Lefebvre, *The Urban Revolution*.

4 Ibid., 81.

5 Ibid. For additional primary materials, analysis, and critique, see Stanek, *Henri Lefebvre on Space*.

6 Marcus and Sarkissian, *Housing as If People Mattered*. The book offers more than 250 concrete guidelines elaborating a broad range of design and management concerns about a variety of types of multifamily housing. The guidelines included photographs, diagrams, and abundant built examples. Marcus and Sarkissian

drew from existing but often overlooked post-occupancy evaluations (POEs), emerging social science research, as well as their own experiences and site visits to different housing communities.

7 Chase, Crawford, and Kaliski, eds., *Everyday Urbanism.*

8 Till, "New Urbanism and Nature."

9 Urban Ecosystems Symposium, University of Minnesota, 25 January 2010.

10 Portney, *Taking Sustainable Cities Seriously.*

11 World Commission on Environment and Development, *Our Common Future.*

12 Hall, *Cities of Tomorrow*; Portney, *Taking Sustainable Cities Seriously*; Harvey, "Possible Urban Worlds."

13 City of Minneapolis, "Sustainability," http://www2.minneapolismn.gov/sustainability/index.htm.

14 City of Minneapolis, Minneapolis City Council, Resolution 2003R-133, adopted April 2003.

15 City of Minneapolis, "Urban Agriculture Policy Plan," 4, https://www2.minneapolismn.gov/government/programs-initiatives/homegrown-minneapolis/zoning-urban-agriculture/.

16 Definitions according to the 2012 City of Minneapolis zoning text amendments, Code 520.160: *aquaculture*: the cultivation, maintenance, and harvesting of aquatic species; *hydroponics*: the growing of food or ornamental crops, in a water and fertilizer solution containing the necessary nutrients for plant growth; *aquaponics*: the combination of aquaculture and hydroponics to grow food or ornamental crops and aquatic species together in a recirculating system without any discharge or exchange of water. Available at https://library.municode.com/mn/minneapolis/codes/code_of_ordinances?nodeId=MICOOR_TIT20ZOCO_CH520INPR_520.160DE.

17 The following definitions are taken from the 2012 Minneapolis zoning text amendments, Code 520.160, ibid.: an *arbour* is "a landscape structure consisting of an open frame with horizontal and/or vertical latticework often used as a support for climbing food or ornamental crops ... [It] may be freestanding or attached to another structure"; a *cold frame* is "an unheated outdoor structure built close to the ground, typically consisting of, but not limited to, a wooden or concrete frame and a top of glass or clear plastic, used for protecting seedlings and plants from cold weather"; *composting* is "the natural degradation of organic material, such as yard and food waste, into soil"; and a *hoop house* is "a temporary or permanent structure typically made of, but not limited to, piping or other material covered with translucent material for the purposes of growing food or ornamental crops ... [It is] considered more temporary than a greenhouse."

18 City of Minneapolis, "Community and Neighborhoods," at https://www2.minneapolismn.gov/resident-services/neighborhoods/.

19 See, for example, MetroBlooms prescriptive literatures and websites, at http://
www.metroblooms.org.

20 See, for instance, Heynen, "Green Urban Political Ecologies."

21 As Sarah, a participant in Northeast Minneapolis, told me, "The rain garden project the neighborhood is trying to do. What is that? I mean, it's the big polluters
who have screwed up our water quality, not an individual's garden. I know the
intentions are good, but it's mismatched effort. We should be regulating big ag!"

CHAPTER TWO

1 The Master Gardener program is an extension program run by the University of
Minnesota in conjunction with Hennepin County. Master gardeners take classes
and participate in ongoing educational activities, while providing a certain number of volunteer hours each season. It is a popular program with retirees and older
residents. Master gardeners live disproportionately in suburban areas, and several
participants reported master gardener activities happening more frequently in
far-flung suburbs. This was one of the reasons Barb, who lives on the North Side,
decided to leave the program.

2 Hitchings, "How Awkward Encounters Could Influence the Future Form of Many
Gardens."

3 In a poem about domestic life and parenthood, "Building My Boat from Kindling," poet Aimee Norton describes the work of laundry as making the necessary
empty brightness of clean clothes.

CHAPTER THREE

1 Gratitude to authors and photographers experimenting with similar questions,
such as the following works, which informed this project: Ahmed, *Living a Fem-
inist Life*; Berlant and Stewart, *The Hundreds*; Crouch, *The Art of Allotments*; de
Leeuw, *Unmarked*; Green, *Atmospherics*; Hartman, *Wayward Lives, Beautiful
Experiments*; Miller, *Dialectograms*; Morris, *The Home Place*; Owens, *Suburbia*;
Sajovic and Butler, *Home from Home*; Salvesen, *New Topographics*; Schwartz,
Waucoma Twilight; Stewart, *Ordinary Affects* and *A Space on the Side of the Road*;
Stoner, *Toward a Minor Architecture*; Temkin, *Private Places*; Tsing, *The Mush-
room at the End of the World*; Tuan, *Space and Place*; Wright, *Casting Deep Shade*.

2 Gibson, Rose, and Fincher, *Manifesto for Living in the Anthropocene*, iii.

3 Ingold, *Being Alive*, 60.

4 Stewart, "Atmospheric Attunements," 445.

5 Ingold, *Being Alive*, 87–8.

6 Deakin, *Notes from Walnut Tree Farm*, 165.

7 Ingold, *Being Alive*, 88.

8 Ibid., 85.

9 Rich, "A Valediction Forbidding Mourning."
10 Ingold, *Being Alive*, 85.

CHAPTER FOUR

1 Geographer Nick Blomley writes, "Rather than merely the objects of ownership, property must be thought of as an organized set of relations between people with regard to a valued resource. As such, it provides a crucial grammar for many of the most consequential relations of social and political life": "The Territory of Property," 593. See also Blomley, "Remember Property?"

2 For a generative introduction to some of these ideas, see Gibson-Graham, Cameron, and Healy, *Take Back the Economy*.

3 For a review, see Hess, "Constructing a New Research Agenda for Cultural Commons." See also Garcia Lopez, Lang, and Singh, "Commons, Commoning and Co-Becomings," for emerging directions in commons research.

4 Gidwani and Baviskar, "Urban Commons."

5 Ibid., 43.

6 See De Angelis, *The Beginning of History*; Chatterton, "Seeking the Urban Common"; Gibson-Graham, *A Postcapitalist Politics*; Harney and Moten, *The Undercommons*; and Hodkinson, "The New *Urban* Enclosures."

7 Consider, for example, Linebaugh's manifesto with respect to recognizing already existing commons all around us, drawn from his historical account of Magna Carta, *Magna Carta Manifesto*. Linebaugh traces past and present commons, and draws on these to argue for the political potential of *commoning* – that is, practices to preserve, maintain, and establish commons of all varieties. Likewise, see Huron's analysis of the ongoing and active making required to establish and maintain housing commons in urban areas, *Carving Out the Commons*.

8 Lefebvre, *Critique of Everyday Life*, vols. 1–3; de Certeau, *The Practice of Everyday Life*, vols. 1 and 2; Amin and Thrift, *Cities*; Goonewardena et al., eds., *Space, Difference, Everyday Life*.

9 Lefebvre, *Rhythmanalysis*.

10 Gidwani and Baviskar, "Urban Commons."

11 Blomley, "Flowers in the Bathtub."

12 The City of Minneapolis enacted an innovative and unique program in 1990, the Neighborhood Revitalization Program (NRP). This channelled funds more directly to neighbourhood organizations, to be allotted to public services and projects identified as priorities by those neighbourhood organizations, with community involvement. It was seen as a novel way to do "planning from below." For background and further reading, see Martin and Pentel, "What the Neighbors Want."

13 Gibson, Rose, and Fincher, *Manifesto for Living in the Anthropocene*, ii.

14 Blomley writes, "It is interesting to speculate on why the territory of property has not received the attention that it deserves. Perhaps its very ubiquity renders it invisible. For legal scholars the tendency has been to bracket territory, given a concern at reducing property to what appears to be an inert space. Property is relations, not spaces, they insist. For geographers, perhaps, the tendency has been to bracket property, or to treat it as a purely economic variable": "The Territory of Property," 605.

CONCLUSIONS

1 Gibson, Rose, and Fincher, *Manifesto for Living in the Anthropocene.*

2 New terrains of experimentation and engagement with environmental change continue to layer and transform our collective understandings. Collaborations across approaches spark new modes of address and representation, such as those assembled in Tsing et al., eds., *Arts of Living on a Damaged Planet.*

3 City of Minneapolis 2040 Plan, https://minneapolis2040.com/. See especially the updated zoning rules related to building forms in effect beginning 1 January 2021, https://minneapolis2040.com/implementation/built-form-regulations/.

BIBLIOGRAPHY

Ahmed, Sara. *Living a Feminist Life*. Durham, NC: Duke University Press, 2017.
- *Queer Phenomenology: Orientations, Objects, Others*. Durham, NC: Duke University Press, 2017.
Amin, Ash, and Nigel Thrift. *Cities: Reimagining the Urban*. Malden, MA: Polity Press, 2002.
Archer, John. *Architecture and Suburbia: From English Villa to American Dream House, 1690–2000*. Minneapolis: University of Minnesota Press, 2008.
Arnold, Jeanne, and Ursula Lang. "Changing American Home Life: Trends in Domestic Leisure and Storage among Middle-class Families." *Journal of Family and Econ. Issues* 28, no. 1 (2007): 23–48.
Bennett, Jane. *Vibrant Matter: A Political Ecology of Things*. Durham, NC: Duke University Press, 2010.
Berlant, Lauren, and Kathleen Stewart. *The Hundreds*. Durham, NC: Duke University Press, 2019.
Blomeley, Nicholas. "Flowers in the Bathtub: Boundary Crossings at the Public-Private Divide." *Geoforum* 36 (2005): 281–96.
- "Remember Property?" *Progress in Human Geography* 29, no. 2 (2005): 125–7.
- "The Territory of Property." *Progress in Human Geography* 40, no. 5 (2016): 593–609.
- *Unsettling the City: Urban Land and the Politics of Property*. New York: Routledge, 2004.
Blunt, Alison, and Robyn Dowling. *Home*. New York: Routledge, 2006.
Braun, Bruce, and Noel Castree, eds. *Social Nature: Theory, Practice and Politics*. Hoboken, NJ: Wiley-Blackwell, 2001.
Cadena, Marisol de la. *Earth Beings: Ecologies of Practice across Andean Worlds*. Durham, NC: Duke University Press, 2015.
Chase, John, Margaret Crawford, and John Kaliski, eds. *Everyday Urbanism*, 2nd ed. New York: Monacelli Press, 2008.
Chatterton, Paul. "Seeking the Urban Commons." *City* 14, no. 6 (2010): 625–8.
Cheng, Irene, Charles L. Davis II, and Mabel O. Wilson. *Race and Modern Architecture: A Critical History from the Enlightenment to the Present*. Pittsburgh: University of Pittsburgh Press, 2020.
Crouch, David. *The Art of Allotments: Culture and Cultivation*. Nottingham, UK: Five Leaves Publications, 2003.

De Angelis, M. *The Beginning of History: Value Struggles and Global Capital*. London: Pluto Press, 2007.

Deakin, Roger. *Notes from Walnut Tree Farm*. Edited by Alison Hastie and Terence Blacker. New York: Penguin Books, 2009.

de Certeau, M. *The Practice of Everyday Life*. Volume 1. Translated by Steven Rendall. Berkeley: University of California Press, 1984.

de Certeau, Michel, Luce Giard, and Pierre Mayol. *The Practice of Everyday Life*. Volume 2: *Living and Cooking*. Translated by Timothy J. Tomasik. Minneapolis: University of Minnesota Press, 1998.

de Leeuw, Sarah. *Unmarked: Landscapes along Highway 16*. Edmonton, AB: NeWest Press, 2004.

Elwood, Sarah, Victoria Lawson, and Eric Sheppard. "Geographical Relational Poverty Studies." *Progress in Human Geography* 41, no. 6 (2016): 745–65.

Federici, Silvia. *Caliban and the Witch: Women, the Body, and Primitive Accumulation*. Chico, CA: AK Press.

Gandy, Matthew. *Concrete and Clay: Reworking Nature in New York City*. Cambridge, MA: MIT Press, 2002.

Garcia Lopez, Gustavo, Ursula Lang, and Neera Singh. "Commons, Commoning and Co-Becoming: Nurturing Life-in-Common and Post-Capitalist Futures." *Environment and Planning E: Nature and Space*. Forthcoming.

Gibson, Katherine, Deborah Bird Rose, and Ruth Fincher. *Manifesto for Living in the Anthropocene*. Brooklyn: Punctum Books, 2015.

Gibson-Graham, J.K. *A Postcapitalist Politics*. Minneapolis: University of Minnesota Press, 2006.

Gibson-Graham, J.K., Jenny Cameron, and Stephen Healy. *Take Back the Economy: An Ethical Guide for Transforming Our Communities*. Minneapolis: University of Minnesota Press, 2013.

Gidwani, V., and A. Baviskar. "Urban Commons." *Economic and Political Weekly* 46, no. 50 (2011): 42–3.

Girling, Cynthia, and Kenneth Helphand. *Yard-Street-Park: The Design of Suburban Open Space*. New York: Wiley, 1994.

Goetz, Edward G., Brittany Lewis, Anthony Damiano, and Molly Calhoun. *The Diversity of Gentrification: Multiple Forms of Gentrification in Minneapolis and St Paul*. Minneapolis: Center for Urban and Regional Affairs, University of Minnesota, 2019.

Goonewardena, Kanishka, Stefan Kipfer, Richard Milgrom, and Christian Schmid, eds. *Space, Difference, Everyday Life: Reading Henri Lefebvre*. New York: Routledge, 2008.

Grampp, Christopher. *From Yard to Garden: The Domestication of America's Home Grounds*. Chicago: University of Chicago Press, Center for American Places, 2005.

Green, Lohren. *Atmospherics*. Niantic, CT: Quale Press, 2014.

Grosz, Elizabeth. *Architecture from the Outside*. Cambridge, MA: MIT Press, 2001.

Groth, Paul. "Lot, Yard, and Garden: American Distinctions." *Landscape* 30, no. 3 (1990): 29–35.

Gundaker, Grey, and Judith McWillie. *No Space Hidden: The Spirit of African American Yard Work.* Knoxville: University of Tennessee Press.

Hall, Peter. *Cities of Tomorrow: An Intellectual History of Urban Planning and Design in the Twentieth Century.* Malden, MA: Blackwell Publishers, 2002.

Haraway, Donna. *Staying with the Trouble: Making Kin in the Chthulucene.* Durham, NC: Duke University Press, 2016.

Harney, Stefano, and Fred Moten. *The Undercommons: Fugitive Planning and Black Study.* New York: Minor Compositions/Autonomedia, 2013.

Harris, Dianne. *Little White Houses: How the Postwar Home Constructed Race in America.* Minneapolis: University of Minnesota Press, 2012.

Harris, Steven, and Deborah Berke. *Architecture of the Everyday.* Princeton, NJ: Princeton University Press, 1998.

Hartman, Saidiya. *Wayward Lives, Beautiful Experiments.* New York: W.W. Norton, 2019.

Harvey, David. "Possible Urban Worlds." In *Justice, Nature, and the Geography of Difference*, 403–38. Cambridge, MA: Blackwell Publishers, 1996.

Head, Lesley, and Pat Muir. *Backyard: Nature and Culture in Suburban Australia.* Wollongong, AU: University of Wollongong Press, 2007.

Henderson, George L. "What (Else) We Talk about When We Talk about Landscape: For a Return to the Social Imagination." In *Everyday America: Cultural Landscape Studies after J.B. Jackson.* Edited by Chris Wilson and Paul Groth. Berkeley: University of California Press, 2003.

Hern, Matt. *What a City Is For.* Cambridge, MA: MIT Press, 2016.

Hess, Charlotte. "Constructing a New Research Agenda for Cultural Commons." In *Cultural Commons: A New Perspective on the Production and Evolution of Cultures.* Edited by Enrico Bertacchini, Giangiacomo Bravo, Massimo Marrelli, and Walter Santagata. Northampton, MA: Edward Elgar Publishing, 2012.

Heynen, Nik. "Green Urban Political Ecologies: Toward a Better Understanding of Inner-city Environmental Change." *Environment and Planning A: Economy and Space* 38 (2006): 499–516.

Hitchings, Russell. "How Awkward Encounters Could Influence the Future Form of Many Gardens." *Transactions of the Institute of British Geographers* 32, no. 3 (2007): 363–76.

Hodkinson, Stuart. "The New *Urban* Enclosures." *City* 16, no. 5 (2012): 500–18.

Huron, Amanda. *Carving Out the Commons: Tenant Organizing and Housing Cooperatives in Washington, D.C.* Minneapolis: University of Minnesota Press, 2018.

Ingold, Tim. *Being Alive: Essays on Movement, Knowledge and Description.* New York: Routledge, 2011.

– *The Perception of the Environment: Essays in Livelihood, Dwelling, and Skill.* New York: Routledge, 2000.

Jackson, J.B. "The Popular Yard." *Places* 4, no. 3 (1987): 26–32.

Jenkins, Virginia Scott. *The Lawn: A History of an American Obsession*. Washington, DC: Smithsonian Institution Press, 1994.

Krueger, Roger, and David Gibbs, eds. *The Sustainable Development Paradox: Urban Political Economy in the United States and Europe*. New York: Guilford Press, 2007.

Lang, Ursula. "The Common Life of Yards." *Urban Geography* 35, no. 6 (2014): 852–69.

– "Connective Tissues: Everyday Engagements with Urban Yards." *GeoHumanities* 4, no. 1 (2018): 230–48.

– "Cultivating the Sustainable City: Urban Agriculture Policies and Gardening Projects in Minneapolis, MN." *Urban Geography* 35, no. 4 (2014): 477–85.

– "Keep Off the Grass! New Directions for Geographies of Yards and Private Gardens." *Geography Compass* 12, no. 8 (2018). https://doi.org/10.1111/gec3.12397.

– "Yards and Everyday Life in Minneapolis." In *Making Suburbia: New Histories of Everyday America*. Edited by John Archer, Paul Sandul, and Katherine Solomonson, 124–40. Minneapolis: University of Minnesota Press, 2015.

Latour, Bruno. *Reassembling the Social: An Introduction to Actor-Network-Theory*. New York: Oxford University Press, 2007.

Lefebvre, Henri. *Critique of Everyday Life*. Volume 1. Translated by John Moore. 1947; New York: Verso Books, 2008.

– *Critique of Everyday Life*. Volume 2: *Foundations for a Sociology of the Everyday*. Translated by John Moore. 1957; New York: Verso Books, 2008.

– *Critique of Everyday Life*. Volume 3: *From Modernity to Modernism*. Translated by Gregory Elliott. 1961; New York: Verso Books, 2008.

– *Rhythmanalysis: Space, Time and Everyday Life*. Translated by Stuart Elden and Gerald Moore. New York: Continuum, 2004.

– *The Urban Revolution*. 1970; Minneapolis: University of Minnesota Press, 2003.

Le Guin, Ursula K. "Deep in Admiration." In *Late in the Day: Poems, 2010–2014*, vii–ix. Oakland, CA: PM Press, 2016.

Linebaugh, Peter. *The Magna Carta Manifesto: Liberties and Commons for All*. Berkeley: University of California Press, 2008.

Lorimer, Hayden. "Cultural Geography: The Busyness of Being 'More-Than-Representational.'" *Progress in Human Geography* 29, no. 1 (2005): 83–94.

MacKinnon, Danny, and Kate Driscoll Derickson. "From Resilience to Resourcefulness: A Critique of Resilience Policy and Activism." *Progress in Human Geography* 37, no. 2 (2013): 253–70.

Marcus, Clare Cooper, and Wendy Sarkissian. *Housing as If People Mattered: Site Design Guidelines for Medium-Density Family Housing*. Berkeley: University of California Press, 1986.

Martin, Judith, and Paula Pentel. "What the Neighbors Want: The Neighborhood Revitalization Program's First Decade." *Journal of the American Planning Association* 68, no. 4 (2002): 435–49.

Massey, Doreen. "Vocabularies of the Economy." *Soundings: A Journal of Politics and Culture* 54 (Summer 2013): 9–22.

McKittrick, Katherine, ed. *Sylvia Wynter: On Being Human as Praxis*. Durham, NC: Duke University Press, 2015.

Merleau-Ponty, Maurice. *Phenomenology of Perception*. Translated by Colin Smith. 1945; New York: Routledge, 2007.

Miller, Mitch. *Dialectograms: From the Ground Up*. http://www.dialectograms.com/.

Morris, Wright. *The Home Place*. 1948; Lincoln: Bison Books, University of Nebraska Press, 1968.

Norton, Aimee. "Building My Boat from Kindling." *Leviathan* 15, no. 2 (2013): 70–1.

– "No Sin Like Arson." *Really System* 2: *Sly Early Stem* (2014). http://reallysystem.org/issues/two/.

Owens, Bill. *Suburbia*. 1973; New York: Fotofolio, 1999.

Pandian, Anand, and Stuart McLean, eds. *Crumpled Paper Boat: Experiments in Ethnographic Writing*. Durham, NC: Duke University Press, 2017.

Portney, Kent E. *Taking Sustainable Cities Seriously: Economic Development, the Environment, and Quality of Life in American Cities*. Cambridge, MA: MIT Press, 2003.

Rendell, Jane, Barbara Penner, and Iain Borden, eds. *Gender Space Architecture: An Interdisciplinary Introduction*. New York: Routledge, 2000.

Rich, Adrienne. "A Valediction Forbidding Mourning." In *Bedford Anthology of American Literature*. Edited by Susan Belasco and Linck Johnson, Vol. 2: 1344. Boston: Bedford, 2008.

Robbins, Paul. *Lawn People: How Grasses, Weeds, and Chemicals Make Us Who We Are*. Philadelphia: Temple University Press, 2007.

Roelvink, Gerda. *Building Dignified Worlds: Geographies of Collective Action*. Minneapolis: University of Minnesota Press, 2016.

Sajovic, Eva, and Sarah Butler. *Home from Home: Documenting Lives through the Regeneration in Elephant and Castle*. 2010. http://homefromhome-online.com/.

Salvesen, Britt, ed. *New Topographics*. Gottingen, DE: Stiedl Books, 2013.

Schwartz, Dona. *Waucoma Twilight: Generations on the Farm*. Washington, DC: Smithsonian Press, 1992.

Simonsen, Kirsten. "In Quest of a New Humanism: Embodiment, Experience and Phenomenology as Critical Geography." *Progress in Human Geography* 37, no. 1 (2012): 10–26.

Singh, Neera. "Affective Ecologies and Conservation." *Conservation and Society* 16, no. 1 (2018): 1–7.

– "Payments for Ecosystem Services and the Gift Paradigm." *Ecological Economics* 117 (2015): 53–61.

Stanek, Lucasz. *Henri Lefebvre on Space: Architecture, Urban Research and the Production of Theory*. Minneapolis: University of Minnesota Press, 2011.

Stewart, Kathleen. "Atmospheric Attunements." *Environment and Planning D: Society and Space* 29 (2011): 445–53.

– *Ordinary Affects*. Durham, NC: Duke University Press, 2007.

– *A Space on the Side of the Road: Cultural Poetics in an 'Other' America*. Princeton, NJ: Princeton University Press, 1996.

Stoner, Jill. *Toward a Minor Architecture*. Cambridge, MA: MIT Press, 2012.

Strunk, Chris, and Ursula Lang. "Gardening as More Than Urban Agriculture: Perspectives from Smaller Cities on Urban Gardening Policies and Practices." *Case Studies in the Environment* 3 no. 1 (2019): 1–8.

Swyngedouw, Erik. "Impossible 'Sustainability' and the Postpolitical Condition." In *The Sustainable Development Paradox: Urban Political Economy in the United States and Europe*, edited by Rob Krueger and David Gibbs, 13–40. New York: Guilford Press, 2007.

Temkin, Brad. *Private Places: Photographs of Chicago Gardens*. Chicago: University of Chicago Press, Center for American Places, 2005.

Tessyot, Georges. *The American Lawn*. New York: Princeton Architectural Press with the Canadian Centre for Architecture, 1998.

Till, Karen. "New Urbanism and Nature: Green Marketing and the Neotraditional Community." *Urban Geography* 22, no. 3 (2001): 220–48.

Tsing, Anna, Heather Swanson, Elaine Gan, and Nils Bubandt, eds. *Arts of Living on a Damaged Planet*. Minneapolis: University of Minnesota Press, 2017.

Tsing, Anna Lowenhaupt. *The Mushroom at the End of the World: On the Possibility of Life in Capitalist Ruins*. Princeton, NJ: Princeton University Press, 2015.

Tuan, Yi-Fu. *Space and Place: The Perspective of Experience*. Minneapolis: University of Minnesota Press, 1997.

Ukeles, Mierle Laderman. "Maintenance Art Manifesto!" Proposal for an Exhibition "CARE." 1969.

Westmacott, Richard. *African-American Gardens and Yards in the Rural South*. Knoxville: University of Tennessee Press, 1992.

Whatmore, Sarah. *Hybrid Geographies: Natures, Cultures, Spaces*. Washington, DC: Sage Publications, 2002.

White, Damian F., and Chris Wilbert. *Technonatures: Environments, Technologies, Spaces, and Places in the Twenty-first Century*. Waterloo, ON: Wilfrid Laurier University Press, 2009.

Wolch, Jennifer, Jason Byrne, and Joshua Newell. "Urban Green Space, Public Health, and Environmental Justice: The Challenge of Making Cities 'Just Green Enough.'" *Landscape and Urban Planning* 125 (2014): 234–44.

World Commission on Environment and Development. *Our Common Future*. Oxford: Oxford University Press, 1987.

Wright, C.D. *Casting Deep Shade: An Amble Inscribed to Beech Trees and Co.* Port Townsend, WA: Copper Canyon Press, 2019.

INDEX